NW8780.
£12.99.
EDC FOUND/99

teaching secondary
CHEMISTRY

Editor Bob McDuell

WITHDRAWN

RRAY

Titles in this series:

Teaching Secondary Biology   0 7195 7637 7
Teaching Secondary Chemistry   0 7195 7638 5
Teaching Secondary Physics   0 7195 7636 9

Photographs are reproduced courtesy of: **cover** M. Agliolo/SPL; **p.12** Chris Priest/SPL; **p.15** Stevie Grand/SPL; **p.35** Science Museum/Science & Society Picture Library; **p.88** Andrew Lambert; **p.219** Ann Ronan/Image Select.

The publishers have made every effort to contact copyright holders. If any have been overlooked they will make the necessary arrangements at the earliest opportunity.

First published in 2000
by John Murray (Publishers) Ltd
50 Albemarle Street
London WIX 4BD

Illustrations by Tony Jones
Layouts by Amanda Easter

Typeset in 12/13pt Garamond by Wearset, Boldon, Tyne and Wear
Printed and bound in Great Britain by the Alden Group, Oxford

A catalogue entry for this title is available from the British Library

ISBN 0 7195 7638 5

Contents

Contributors

Phil Hyde is a practising teacher, having previously spent some time in academic and industrially funded research. He is Head of Science at a large comprehensive school in Derby specialising in Chemistry and has a wide experience in Science examining, both for domestic and overseas courses. Currently, he is an examiner for KS3 Science and a Team Leader for GCSE Double Science.

Judith Johnston is a practising teacher and has taught mainly Chemistry, 11 to 19, for over 30 years in grammar and secondary schools and in Colleges of Technology. She is Head of Chemistry at Hunterhouse College in Belfast. She has written revision guides for GCSE and A level Chemistry aimed specifically at the Northern Ireland curriculum.

David Lees is Head of Science at Chepstow Comprehensive School. He has been involved with the production and marking of science examinations for the past 20 years, and is now Principal Examiner for GCSE Double Science. In addition to this work in the UK, he is a Principal Examiner for papers taken in many other parts of the world. Originally trained in Biochemistry, he writes both Chemistry and Biology material for examination boards and publishers.

Bob McDuell taught Chemistry in secondary schools for 33 years before retiring. He has been a Chief Examiner for Chemistry and Science from 1974 and also examines at A level and International Baccalaureate. He also examines Chemistry for overseas O level and IGCSE and trains teachers in the UK and throughout the world. He has written over 70 books from primary to A level including many books at GCSE and A level.

Geoff Mines is a Chemistry teacher and Head of the Science Department at Bedford School. He is a Principal Moderator at both GCSE and A level and has led many INSET sessions around the country through the many changes to coursework assessment. He has been involved with Nuffield courses for many years and has also acted as a consultant for QCA, monitoring examination standards.

Ray Oliver combines the roles of Head of Science and education consultant. He is the author of many books and also works as a freelance editor. He writes about all aspects of science education for both teachers and children and has a long-standing interest in education–industry liaison. In recent years he has acted as the co-ordinator for a primary–secondary science cluster group in London.

John Payne is a practising Chemistry and Science teacher. He has taught in an 11-to-18 comprehensive school for over 20 years. He has been involved in many aspects of examination work for 18 years, ranging from CSE, O and A level in the past to GCSE and CoA at present. He is currently Chief Examiner and Principal Coursework Moderator for Science Certificate of Achievement (Science Plus) and is Principal Examiner for GCSE Science (Chemistry). He has been involved in the writing of books including examination practice and study guides.

Gareth Pritchard is a member of the Senior Management Team and teaches Chemistry in a Hertfordshire comprehensive school. His long teaching experience has included many years as a Head of Science. His extensive involvement as an Examiner started in the early 1960s and has included Chief Examiner posts at O level, A level and GCSE. He is presently a Principal Examiner in Chemistry at GCSE.

Neil Rowbotham is now a Deputy Head after 25 years as Head of Chemistry. He is a Principal Examiner for A level Chemistry and was previously an Assistant Chief Moderator for GCSE Coursework. He has been honoured with the Royal Society of Chemistry Award for Chemical Education.

Acknowledgements

The authors and editor are very grateful to the following for their advice during the preparation of this book:

Jackie Callaghan
Harry Chalton
Mike Coles
Alistair Fleming
Ted Lister
Colin Osborne
Doug Peacock
Peter Radford
David Sang

The Association for Science Education acknowledges the generous financial support of ESSO and the Institution of Electrical Engineers (IEE) in this project. Both ESSO and IEE provide a range of resources for science teaching at primary and secondary levels. Full details of these resources are available from:

ESSO Information Service
PO Box 94
Aldershot
GU12 4GJ
(tel: 01252 669663)

IEE Educational Activities
Michael Faraday House
Six Hills Way
Stevenage
SG1 2AY
(e-mail: nsaunders@iee.org.uk)
(web site: www.iee.org.uk/Schools/)

Introduction

Bob McDuell

This book is one of a series of three ASE handbooks, the others being parallel volumes on Biology and Physics. It is hoped that this book will convey to readers the enthusiasm for Chemistry of the author team and enable teachers of 11–16 year olds better to prepare pupils with the necessary groundwork in Chemistry. The author team has kept in mind a teacher confronted with the task of teaching a specific topic, e.g. rocks, in the near future. What does such a teacher need to produce a series of effective lessons? In particular, where teachers are not specialist Chemistry teachers, they will want to be confident about what practical work could be done safely and healthily to support the topic.

Who is the book for?

In writing their chapters, authors have identified a range of likely readers:

- new or less experienced Chemistry teachers – though almost every Chemistry teacher should find much of value
- biologists, physicists and science generalists who find themselves teaching parts of the Chemistry curriculum
- student teachers and their tutors/mentors
- heads of department who need a resource to which to direct their colleagues
- teachers of Biology and Physics who want to know more about the Chemistry curriculum.

While we have taken into account the current UK and international syllabus requirements, we have not stuck closely to any one curriculum. We expect that this book will be appropriate to secondary Chemistry teachers throughout the world.

What should you find in the book?

We expect you, the reader, to find:

- good, sensible, stimulating ideas for teaching Chemistry to 11–16 year olds
- suitable practical work for both pupil experiments and teacher demonstrations

- suggestions for extending the range of approaches and strategies that you can use in your teaching
- things to which pupils respond; things that fascinate them
- confirmation that a lot of what you are doing is fine: there often isn't a single correct way of doing things.

How can you find what you want?

The author team divided secondary Chemistry into nine areas. These correspond to Chapters 1 to 9. It should be clear from the Contents page (page *iii*) what each chapter roughly consists of. A more detailed indication is given by the 'Content map' at the beginning of each chapter. This divides up each chapter into about half a dozen sections. For specific topics, consult the index. This contains the likely terms you might want to look up, e.g. diffusion or enzymes.

What is in each chapter?

Each chapter contains:

- a content map which divides up the chapter into shorter sections
- a set of possible teaching routes through the chapter
- a brief section on what pupils may have learnt or experienced about the topic in their primary science lessons
- an outline teaching sequence showing how concepts can be developed throughout the 11–16 phase
- suggested practical work, with health and safety information, required materials and procedures
- what might be expected in each practical
- information about likely pupil misconceptions
- suggestions for the use of ICT (highlighted by the use of icons in the margin)
- suggestions about the use of books, videos and other resources
- opportunities for investigative work
- links with other areas of Chemistry or Science.

Health and safety first!

A chemistry laboratory is potentially a dangerous place, probably one of the most dangerous in the school. As a teacher you have a duty of care for the health, safety and well-being of the pupils. Despite being such a dangerous place there are relatively few serious accidents. This is due to the work of organisations such as CLEAPSS, SSERC (in Scotland) and ASE in identifying possible hazards and disseminating information, but also to teachers who prepare for practical lessons by considering possible hazards and adopting measures to reduce the risk of these hazards causing harm. Before planning any practical activity:

1. Make sure you are aware of the chemistry of any chemical reactions taking place.
2. Identify possible hazards both in materials and procedures. In doing this HAZCARDS (distributed by CLEAPSS 1995 or 1998 update) are very useful and other publications may sometimes be needed.
   Safety in Science Education. DfEE, HMSO (1996).
   Topics in Safety, 2nd edition. ASE (1988).
   Safeguards in the School Laboratory, 10th edition. ASE (1996).
   CLEAPSS Laboratory Handbook. CLEAPSS (1997 and any later supplements).
   CLEAPSS Shorter Handbook. CLEAPSS (1999 and any later supplements).
   Hazardous Chemicals. A Manual for Science Education. SSERC (1997). A CD ROM version (1998) is also available.
3. Ensure you have carried it out yourself beforehand so it is not happening for the first time with a class.
4. Consult your employer's risk assessment – this may well be included in your department's scheme of work – or CLEAPSS HAZCARDS or another model risk assessment. If your employer has not provided a risk assessment, a special risk assessment may be necessary.
5. Do not modify what you have planned to do.

You should be aware of the safety recommendations from CLEAPSS, SSERC and ASE but you should also check if there are any limitations on practical work made by your employer.

Many accidents can occur through teachers or pupils not wearing eye protection and not using appropriate safety screens. It is sometimes a temptation to remove these so that pupils can see more clearly what is happening, perhaps in a demonstration. Consider possibly carrying out a demonstration several times with smaller groups rather than one with the whole class. This can be particularly useful when a demonstration is taking place in a fume cupboard. It means providing something worthwhile for the others to do when they are not watching the demonstration.

Sometimes pupils are tempted to remove goggles because they are scratched or uncomfortable. They should be checked on a regular basis.

When considering what you are going to do, make a conscious effort to ensure pupils use the lowest possible concentrations of solutions. This reduces the risks but also reduces the costs! Pupils in the lower school using acids and alkalis with simple indicators can use solutions of concentration 0.1 mol/dm^3. However, towards the end of the course, when pupils are working out the energy liberated when acids and alkalis react, the solution may be 2 mol/dm^3. Any accident with sodium hydroxide solution, which is highly caustic, will have less serious consequences if the solution is diluted.

Care should be taken with the use and disposal of highly flammable liquids. Liquids, such as methylated spirit, should not be heated with a naked flame but by immersing a test-tube or boiling tube containing the liquid in a beaker of hot water, with all flames extinguished.

Consider the reaction of potassium or sodium with cold water. This is a common demonstration where many things can go wrong. There is a danger in being encouraged to use a piece larger than recommended. If the experiment is repeated with further pieces of potassium or sodium, it will be at a higher temperature and so the reaction is faster. The potassium or sodium is inclined to spit. Accidents could occur if appropriate safety screens are not used. Also if the lid is not replaced on the bottle of potassium or sodium a spark could catch the paraffin oil alight.

It is a good idea to make pupils aware of dangers from the start and it is a requirement of the National Curriculum. This should include the necessity to wear eye protection, following instructions precisely and what to do if an accident occurs. At certain stages during the course they should be encouraged to do a risk assessment for themselves. A photocopiable example of an appropriate form is provided on page 300, with a completed example also supplied. Doing this you are encouraging them to work safely.

In the event of an accident it is important for you to know the correct procedures for first aid and emergency treatment. Within the staff of the school there should be at least one person trained in first aid and provided with a fully stocked first-aid box. Very serious accidents involving loss of limbs, damage to eyes and unconsciousness have to be reported to the Health and Safety Executive. There should be a system of recording even minor accidents, such as small burns and cuts, in case reference needs to be made to what happened and what was done later.

Useful contacts
CLEAPSS School Science Service, Brunel University, Uxbridge, Middlesex UB8 3PH (tel: 01995 251496; fax: 01895 814372; e-mail: science@cleapss.org.uk; web site: www.cleapss.org.uk).
Most local authorities outside Scotland have subscribed to this service, and schools are therefore able to use their services free of charge, as are the 2000+ independent schools, colleges, teacher training establishments and overseas institutions which subscribe directly.

SSERC (Scottish Schools Equipment Research Centre). This has a similar role to CLEAPSS, in Scotland. St Mary's Buildings, 23 Holyrood Rd, Edinburgh EH8 8AE (tel: 0131 558 8180; fax: 0131 558 8191; e-mail: sserc@mhie.ar.uk; web site: www.sutc.org.uk/resources/sserc).

Association for Science Education, College Rd, Hatfield, Hertfordshire AL10 9AA (tel: 01707 283000; fax: 01707 266532; e-mail: ase@asehq.telme.com; web site: www.ase.org.uk).

A consultant to CLEAPSS has read this text and confirms that, in the draft checked, the identification of hazards and the precautions given either conform with published general risk assessments or, if these are not available, are judged to be satisfactory.

1 *Introducing chemistry*

Gareth Pritchard

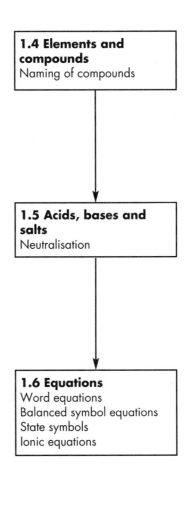

1.1 Solubility
Solubility in water and
  non-aqueous solvents
The effect of temperature on
  solubility
Saturated solutions

**1.2 Separation
techniques**
Filtration
Centrifuging
Simple distillation
Fractional distillation
Chromatography
Sublimation

1.3 Hardness in water
Methods of softening water

**1.4 Elements and
compounds**
Naming of compounds

**1.5 Acids, bases and
salts**
Neutralisation

1.6 Equations
Word equations
Balanced symbol equations
State symbols
Ionic equations

◆ *Choosing a route*

In everyday chemistry at home, at work or in leisure activities, it is important that we can depend on the purity of substances. For example, we are used to seeing the ingredients printed on the sides of cereal boxes or yoghurt tubs to tell us what we are eating. In the treatment of illnesses, drugs may be prescribed and their manufacturing process has to be scrupulously supervised so that pure dependable products are made. Obviously, there are additional costs in producing very pure chemicals.

Pure elements and compounds have particular characteristics, which a chemist calls their 'properties'. These include their melting and boiling points, their physical appearance, their chemical reactions with particular chemicals and their solubility in chosen solvents.

In a chemical process, manufacturers will have ensured that the substances they react together (the reactants) and the substances they produce (the products) are as pure as possible, and nearly all the substances which are used will have gone through two stages:

- separation from a mixture
- testing to validate their purity.

The actual quantities of substances used will also have been measured quantitatively to make sure the reactions 'go' efficiently.

In this section there are two strands, which can be related or dealt with separately:

1.1 Solubility	1.4 Elements and compounds
1.2 Separation techniques	1.5 Acids, bases and salts
1.3 Hardness in water	1.6 Equations

Many of the topics in this section are at a relatively low level. However, in external examinations pupils often show a lack of basic understanding, especially of elements and compound formation.

Also, pupils should be able to write chemical equations. At one time equations were taught early and candidates became proficient with practice over two or three years. There is a tendency today to leave equations till later and rely on pupils' greater maturity.

Pupils who are intending to continue further studies in Chemistry post 16 should be able to write chemical equations using symbols and balance them. All pupils should be able to interpret simple symbol equations, for example be able to write down the names of the products in an equation such as

$$CuO + H_2SO_4 \rightarrow CuSO_4 + H_2O$$

1.1 Solubility

◆ *Previous knowledge and experience*

In primary school Science, pupils may well have used everyday substances like salt or sugar and shaken them up with water in jars. They may know the difference between the words 'soluble' and 'insoluble', or have met the words 'dissolving' and 'solution'. Pupils should be aware that there is a limit to the mass of solid that can dissolve in a given amount of water. This limit is different for different solids and often varies with temperature.

◆ *A teaching sequence*

Pupils will need to learn the following terms:

- solute – the substance being dissolved
- solvent – the substance in which the solute is dissolved.

In this work pupils should learn not to confuse the terms 'reacting' and 'dissolving'.
  The process of dissolving can involve two mechanisms:

- mixing of particles of the solute with particles of the solvent because they have similar (covalent) bonding between their particles, or
- the separation of the compound into ions, as takes place when water is the solvent with ionic compounds.

If the solvent is suitably removed, e.g. by evaporation, the original solute is left behind.
  'Reacting' can look like dissolving, e.g. adding sodium carbonate to a dilute acid gives a lot of gas bubbles (effervescence) and a colourless solution. This solution looks the same as a solution produced by dissolving sodium carbonate in water. Sodium carbonate can be recovered from a solution of sodium carbonate by evaporation; sodium carbonate cannot be recovered by evaporation of the solution formed when sodium carbonate reacts.

Solubility in water and non-aqueous solvents

Materials
- eye protection
- test-tubes
- test-tube rack
- cotton buds
- glass rod
- nail-varnish applicator
- propanone (acetone)
- spatula
- microscope slide
- nail varnish
- sodium chloride (salt)

Safety
- *Propanone and nail varnish are volatile, highly flammable and smelly. Use only in a well-ventilated room or a fume cupboard, with no naked flames near them.*
- *Keep bottles of propanone and nail varnish stoppered when not in use.*
- *Wear eye protection when using propanone.*

Procedure 1
1. Put enough sodium chloride into a test-tube to half-fill the curved base.
2. Add (tap) water to quarter-fill the test-tube.
3. Shake thoroughly.

What you might expect
The salt should dissolve to give a colourless solution.

Procedure 2
1. Put two or three grains of salt on to a microscope slide.
2. Add three or four drops of propanone.
3. Stir gently with the glass rod.

What you might expect
The salt will not dissolve. The propanone will evaporate, leaving the salt crystals behind.

Procedure 3
1. Put a small dab of nail varnish on to a microscope slide and allow it to dry.
2. Wet a cotton bud with water and rub it over the dried nail varnish.

What you might expect
The nail varnish should not dissolve.

Procedure 4
1. Put a small dab of nail varnish on to a microscope slide and allow it to dry.
2. Wet a cotton bud with propanone and rub it over the dried nail varnish.

What you might expect
The nail varnish should start to dissolve. Its colour should be seen on the cotton bud.

These experiments illustrate the point that sodium chloride dissolves in a (polar) solvent like water. Nail varnish does not dissolve in a polar solvent but does dissolve in a (non-polar) solvent like propanone.

The effect of temperature on solubility

Lead(II) chloride crystals are insoluble in cold water but dissolve in hot water. The experiment needs some care and may be better as a demonstration. The crystals are first prepared, then separated and the action of heat investigated.

Materials
- eye protection
- test-tubes
- spatula
- Bunsen burner, clamp and stand
- filter funnel and paper, or centrifuge
- hydrochloric acid (2 mol/dm^3)
- lead(II) nitrate solution (0.02 mol/dm^3)
- sodium iodide solution (0.04 mol/dm^3)

Safety
- *2 mol/dm^3 hydrochloric acid is an irritant.*
- *Warming the lead(II) chloride crystals in water must be done gently. Heating too vigorously can cause the crystals to be ejected from the test-tube.*
- *Hands need to be washed thoroughly after handling lead(II) chloride crystals.*
- *Wear eye protection.*

<u>Procedure</u>
1. Pour 3 cm depth of lead(II) nitrate solution into a test-tube.
2. Half-fill the test-tube with hydrochloric acid.
3. Let the white precipitate settle. (This could be helped by filtering it or by using a centrifuge.)
4. Collect a small portion of the solid lead(II) chloride crystals and put it into a clean test-tube.
5. Half-fill the test-tube with cold water and set up the apparatus shown in Figure 1.1.
6. Wearing eye protection, warm the test-tube with the Bunsen burner until the liquid boils.

Figure 1.1
Crystals of lead(II) chloride will dissolve in water on heating.

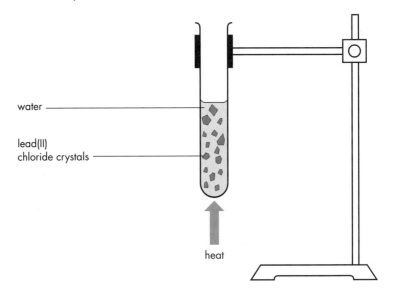

water

lead(II) chloride crystals

heat

<u>What you might expect</u>
The crystals will dissolve as the water approaches boiling point. If the test-tube is clamped vertically in a retort stand and left undisturbed while it cools, sparkling crystals will form.

An alternative recipe is to use lead(II) nitrate and sodium iodide solutions. The lead(II) iodide crystals formed are yellow.

Saturated solutions
Pupils extend their understanding of saturation by studying the topic quantitatively. The solubility of a solute in a particular solvent is the mass of solute which saturates 100 g of solvent at a given temperature.

The solubility of most solid solutes increases as the temperature rises. One can consider this from simple kinetics by suggesting that the higher the temperature, then the more energy there is available for the breaking of bonds in the crystals. This enables the solvent action of the water to be more effective, taking the separating ions into solution. One well-known exception is sodium chloride: its solubility in water is almost constant at 36 g per 100 g of water up to 100 °C.

However, gases become *less* soluble as the temperature rises: the heat energy gives more kinetic energy to the gas molecules, allowing them to escape from the solution. There is more oxygen dissolved in cold water than in warm water – an important fact for fish life. Salmon, for instance, cannot survive if the water temperature is higher than about 15 °C; there is insufficient oxygen dissolved in the water above this temperature. Fish caught in rivers generally do not survive in fish tanks indoors, where the water is warmer and hence deficient in dissolved oxygen.

Subjecting a gas to higher pressure will cause more of it to dissolve. When the pressure is released, the gas bubbles out again. This principle is used in the making of fizzy drinks, in which carbon dioxide is the dissolved gas. When the ring-pull on a can of fizzy drink is pulled off, the carbon dioxide starts to bubble out and the drink eventually goes 'flat'.

Pupils may be asked to interpret simple solubility curves, such as those in Figure 1.2. These plot the solubility in grams per 100 g of water against temperature. From them one can read off the solubility at different temperatures and hence work out how much of a given solute will crystallise out when a solution is cooled.

Figure 1.2
Solubility curves.

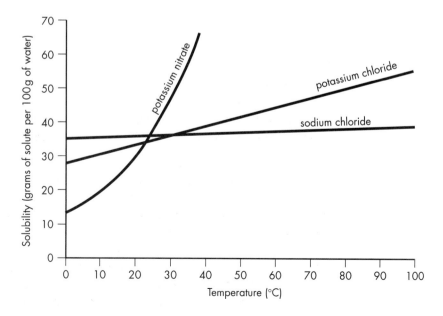

It should be noted that when solutions of laboratory reagents are prepared, their concentration is normally quoted in either grams per litre of solution or moles per litre of solution (1 litre is the same as 1 dm^3, see Chapter 5).

When an ionic substance dissolves in water, the water molecules become somewhat loosely attached to the separated ions. At higher levels, pupils will be expected to write ionic equations which involve 'state symbols'. In the example of sodium hydroxide, the dissolved ions would be written as Na$^+$(aq) and OH$^-$(aq), the 'aq' representing the attached water molecules.

◆ *Further activities*

- ◆ Demonstration of the 'fountain' experiment (No. 79 in *Classic Chemistry Demonstrations*, Royal Society of Chemistry, 1998, ISBN 1 870343 38 7). This can be used to demonstrate the high solubility of ammonia, hydrogen chloride and sulphur dioxide in water.
- ◆ Demonstration to show light scattering in a colloidal solution (No. 82 in *Classic Chemistry Demonstrations*, Royal Society of Chemistry, 1998, ISBN 1 870343 38 7).

◆ *Enhancement ideas*

- ◆ Pupils can use data from a good data book to plot solubility curves using either a graph drawing package or graph paper. Suitable data books include:

  Stark, J.G. & Wallace, H.G. (1999). *Chemistry Data Book*, 2nd edition in SI. London, John Murray.
  Revised Nuffield Advanced Science (1984). *Book of Data*. Harlow, Longman.

1.2 Separation techniques

◆ *Previous knowledge and experience*

In primary school Science, pupils will probably have met the techniques of filtering, sieving and decanting, and they may have tried to separate the colours in, say, ink using blotting paper or filter paper. They may also have tried separating iron from a mixture with sulphur using a magnet.

From primary school, pupils should be aware of the scientific terms 'evaporating' and 'condensation'. For example, pupils may be familiar with the evaporation of water puddles on a warm pavement after a summer shower, or condensation on a cold window overnight. Evaporation that takes place at the boiling point is called 'boiling'. It is unlikely that primary pupils will have met the combined technique of evaporation and condensation: this is known as 'distillation'.

◆ *A teaching sequence*

It is important for pupils to realise that the elements in a *compound* cannot be separated by physical means: chemical reactions must be used. It is only the components of a *mixture* which can be separated by physical means. To separate substances from each other, there must be some difference in properties between them. This could be, for example, a difference in particle size, in solubility, in boiling point or in density.

Table 1.1 *Separation techniques.*

Difference in property	Technique which could be used
Particle size	Filtration
Density	Separating funnel, centrifuge or decanting
Boiling point	Distillation
Solubility	Chromatography
Immiscible liquids	Separating funnel

Filtration to separate sand from salt

This experiment uses the fact that salt dissolves in water (is soluble) and sand does not (is insoluble) to bring about a separation.

Materials
- eye protection
- filter funnel, paper and stand
- glass rod
- 100 cm^3 plastic beaker
- 250 cm^3 glass beaker
- tripod, gauze and heatproof mat
- Bunsen burner
- evaporating basin
- mixture of sand and salt (pre-mix a level dessertspoonful of each)

Safety
- *Wear eye protection when stirring mixtures and when using a Bunsen burner.*

Procedure
1. Pour the sand and salt mixture into the plastic beaker.
2. Fill the beaker to the 25 cm^3 mark with water.
3. Stir the mixture for about 2 minutes to dissolve all the salt.
4. Filter the mixture into the evaporating basin. (The filtrate is salt solution; the residue is sand.)
5. Put the evaporating basin over a beaker of water as in Figure 1.3 and warm the water until it boils; then adjust the heat so that it simmers gently.
6. Watch for a tide mark appearing around the meniscus in the evaporating basin. This shows that the solution is saturated and means that if the solution is now cooled, salt crystals will form, as tiny cubes. Beware of 'spitting' if the solution is allowed to evaporate too much.

Figure 1.3
Using a water bath to concentrate the salt solution.

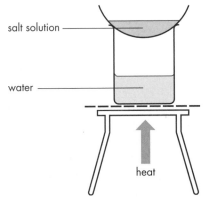

salt solution

water

heat

There are some useful techniques which pupils should master in this experiment.

- Proper use of filter paper – choose a medium grade because the sand particles are bulky. Make certain that the paper is properly folded (i.e. into halves and then into quarters), so that it makes a cone inside the filter funnel. Wet the inside of the filter paper with a little water before filtering starts; this keeps the filter paper in position.
- Pouring liquids – use a glass rod as a guide, held over the top of the beaker as one would hold a pen (see Figure 1.4). The liquid can then be directed into the centre of the filter paper without spilling it. Fill the cone to within 1 cm of the top – anything higher can seep over the top. During filtering, do not touch the paper with the glass rod – it might tear it.

Figure 1.4

Using a glass rod to guide liquid into a filter funnel.

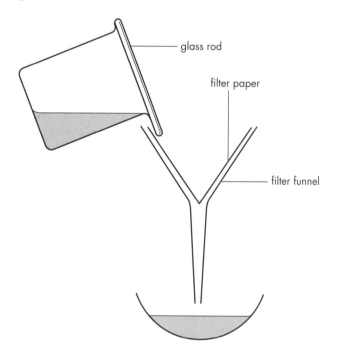

glass rod

filter paper

filter funnel

- Concentrating the solution – if it is heated directly over a Bunsen burner, it evaporates too quickly and the crystals are misshapen. The use of a beaker as a water bath guards against this. An alternative way of testing whether the crystallisation point has been reached is to dip the glass rod into the hot solution to get a drop on the end, blow it gently and see if it goes opaque (because crystals are starting to form on the cooled rod).

Centrifuging

This technique is usually used to separate a solid from a liquid. Pupils may compare the centrifuge with a spin-drier, where the rapidly spinning drum causes the water to be thrown outwards.

Centrifuging is used commercially to separate cream from milk, for example, or blood products from whole blood to produce plasma, Factor VIII and platelets, etc. (see Figure 1.5).

Figure 1.5

Using a centrifuge to separate components in the blood.

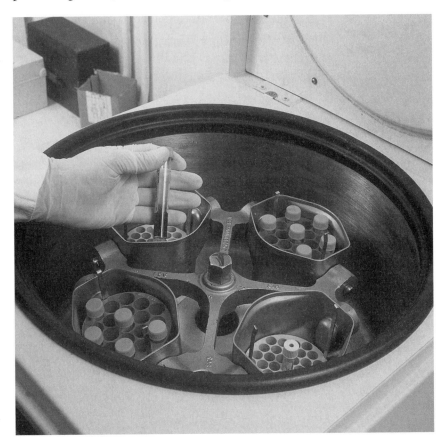

In the laboratory, it can be demonstrated by putting a suitable mixture into the left-hand tube of the centrifuge and balancing it with water to the same depth in the right-hand tube. (If the counterbalancing tube is not used, there is a risk of damage to the centrifuge.) The denser component sinks to the bottom, where it is compacted. The supernatant liquid can be removed using a pipette or by decanting it off.

A suitable sample can be made by mixing together equal volumes of lead(II) nitrate and sodium carbonate solutions. Insoluble lead(II) carbonate will be compacted; the supernatant liquid is sodium nitrate solution.

Simple distillation

Distillation is a two-stage process – evaporating at boiling point followed by condensation. It is used as a way of separating and recovering a solvent from a solute. In its natural state, the solute is a solid. If two or more liquids of different boiling points are to be separated, fractional distillation is used instead; this technique is much practised in oil refining (see page 17).

Distillation of salt solution

This experiment is probably best done as a demonstration. As a preliminary, pupils could carry out a similar experiment without a condenser to collect some of the water in blue ink (Figure 1.6).

Figure 1.6
Simple distillation apparatus (without a condenser).

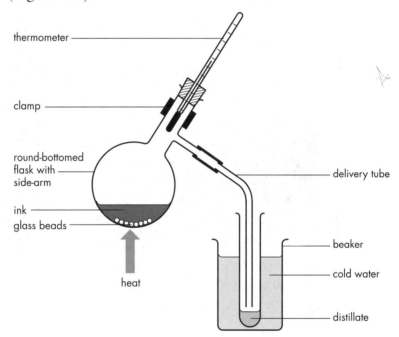

Materials
- eye protection
- distillation flask
- glass beads
- condenser
- thermometer, −10 to +110 °C range
- tripod, gauze and heatproof mat
- Bunsen burner
- beaker
- sodium chloride (salt) solution
- cobalt chloride paper

<u>Safety</u>
- *Check that the tubing is not kinked and that the outlet tubing easily reaches a sink.*
- *Wear eye protection when carrying out distillation.*

<u>Procedure</u>
1. Set up the apparatus as shown in Figure 1.7. Note that:
    - the distillation flask should be no more than one-third-filled with the mixture to be distilled
    - the thermometer bulb should be opposite the side-arm to give the boiling point of the distillate
    - the inlet tubing from the water tap must be connected to the *bottom* entrance of the condenser, so that the incoming water pushes out all the air from within it
    - the beaker must be suitably placed under the condenser exit to catch the distillate
    - the distillation flask must contain eight to ten glass beads, to prevent the boiling becoming too vigorous.

Figure 1.7
Distillation apparatus.

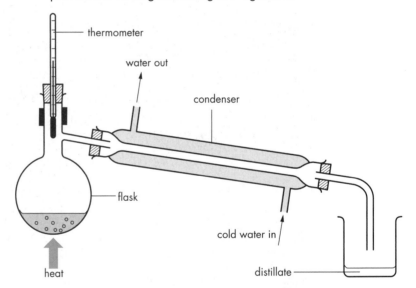

2. Heat the mixture with the Bunsen burner and note the temperature at which condensation first appears. This should be 100 °C, but if atmospheric pressure is low, the boiling point will be lower.
3. Collect a sample of the colourless distillate in the beaker. Do not let the distillation flask boil dry – it will crack.
4. The liquid can be tested to show the presence of water by touching it with some dried cobalt chloride paper; this is blue when dry and turns feebly pink (virtually colourless) when water is added to it. (The criterion for assessing whether a colourless liquid contains *only* water is that its boiling point is exactly 100 °C at 1 atmosphere pressure.)

Fractional distillation

In oil refining, fractional distillation separates the different liquid components of similar boiling points in crude oil. Demand outstrips supply, however, and further treatment involving 'cracking' and 're-forming' is needed to obtain the required 'popular' components such as petrol (see Chapter 3).

Table 1.2 *The fractions obtained from crude oil.*

Boiling point range (°C)	Fraction	Example of use
Below room temperature	Refinery gas	'Camping gas'
30–160	Gasoline	Petrol
160–250	Kerosine	Aviation fuel
250–350	Diesel oil	Fuel for trains, lorries, etc.
Above 350	Bitumen	Roads, roofing

Figure 1.8
Fractional distillation of crude oil in an oil refinery.

Industrially, the separation is achieved by having outlet pipes at different heights on the column (see Figure 1.8). Each height corresponds to the boiling point range of a particular component: the components with the lowest boiling points will travel furthest up the column.

In the laboratory a vertical column packed with glass beads, or with a specially designed interior to give plenty of surface area on which condensation can occur, can be set up as in Figure 1.9. For the following experiment, however, a simple distillation apparatus is adequate.

Figure 1.9
Fractional distillation.

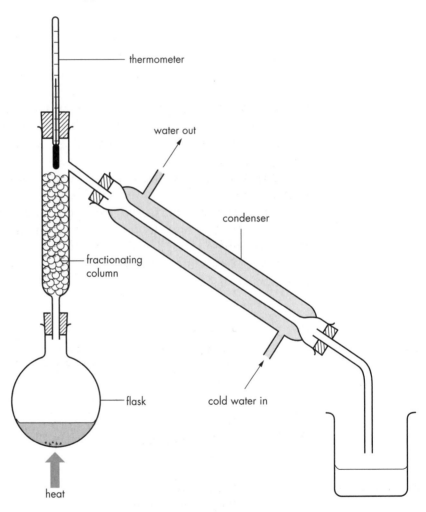

thermometer

water out

condenser

fractionating column

flask

cold water in

heat

Separation of a mixture of liquids by fractional distillation
The principle of fractional distillation can be demonstrated as a class exercise in the laboratory using a mixture of water and the dye fluorescein.

Procedure
1. Set up the apparatus as for simple distillation (see Figure 1.7).
2. One-third-fill the distillation flask with the water containing just sufficient dye to make it visibly green. Include a few anti-bumping glass beads.
3. Distil as before and collect the distillate.

What you might expect
It will boil at 100 °C and leave the green dye behind. This implies that the dye has a higher boiling point than the water.

Demonstration of fractional distillation of 'crude oil'

It is no longer possible to use real crude oil in schools. It contains up to 5% benzene, and benzene can produce anaemia, leukaemia and other blood disorders. Higher boiling point fractions contain compounds with four to six aromatic rings, which are potent carcinogens. Crude oil also contains many sulphur compounds which can give rise to disagreeable odours during distillation.

It is possible to prepare a less hazardous alternative using aliphatic hydrocarbons, however. Table 1.3 gives two recipes. The volumes given make sufficient 'crude oil' for 16 distillation experiments. This 'crude oil' can be given an appropriate black colour by adding powdered charcoal or one small spatula measure of black oil paint. (Windsor and Newton Ivory Black in a tube works well.)

Table 1.3 *Recipes for substitute 'crude oil'.*

Constituent	Boiling point range (°C)	Recipe 1 (volumes in cm³)	Recipe 2 (volumes in cm³)
Petroleum ethers	60–80	6	4
	80–100	4	4
	100–120	4	4
White spirit	120–210 (the 150–180 range produces the most distillate)	11	11
Paraffin oil (kerosine)	120–240 (the 180–210 range produces the most distillate)	20	15
Lotoxane*	185–215	—	10
Paraffin liquid (medicinal paraffin)	250 and over	55	50

*A hydrocarbon solvent available from Griffin and George (Fisons Scientific Equipment catalogue no. L/2550/17).

Alternatively, this experiment could be carried out by a set of Year 9 pupils with close supervision.

This experiment needs to be done slowly to get good results.

Materials
- eye protection
- side-arm boiling tube
- thermometer, 0–360 °C
- cork to fit boiling tube, with one hole
- piece of rubber tubing
- right-angled delivery tube
- 6 ignition tubes
- Bunsen burner
- stand, boss and clamp
- 1 cm³ graduated pipette
- mineral wool
- substitute 'crude oil' (see Table 1.3)

Safety
- *All petroleum ethers are highly flammable.*
- *60–80 °C petroleum ether is harmful.*
- *The experiment must be carried out in a well-ventilated room.*
- *Wear eye protection when carrying out fractional distillation.*

Procedure
1. Put a piece of mineral wool in the side-arm boiling tube and add 6 cm³ of substitute 'crude oil'.
2. Clamp the apparatus as shown in Figure 1.10, making sure that the bulb of the thermometer is level with or just below the level of the side-arm. Keep the piece of rubber tubing between the side-arm boiling tube and the delivery tube as short as possible to reduce condensation of the distillate to a minimum.

Figure 1.10
Fractional distillation of 'crude oil'.

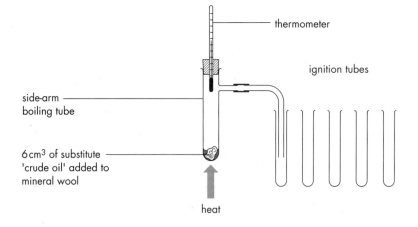

thermometer

ignition tubes

side-arm boiling tube

6 cm³ of substitute 'crude oil' added to mineral wool

heat

3. Warm the boiling tube very gently up to 100 °C, collecting the distillate in the first ignition tube.
4. Place the delivery tube into the second ignition tube and heat the boiling tube up to 150 °C.
5. Repeat step 4, but heat up to 200 °C.
6. Repeat step 4, but heat up to 250 °C.
7. Repeat step 4, but heat up to 330 °C.
8. Using the graduated pipette, place 1 cm³ of water in an ignition tube. Measure the height of the water in the ignition tube. Use this to estimate the volumes of the different fractions collected and to find the percentage of the original oil not distilled.

Sample results

Boiling point of fraction (°C)	Volume (cm³)
Up to 100	0.8
100–150	0.8
150–200	1.6
200–250	0.6
250–330	0.4

Chromatography

This technique is used to separate very small quantities of substances which vary in solubility in different solvents. Modern methods can involve the use of thin absorbent films on microscope slides ('thin layer chromatography') or columns through which a gas is passed as a solvent ('gas chromatography'). However, it is quite sufficient to use paper as the medium on which the separation is effected.

The principle behind the technique is that if a solute is given a 'choice' of two solvents in which to dissolve, it will distribute itself between the two, the proportions depending on how soluble it is in each of them. If a drop of sample is placed on a piece of chromatography paper and then the paper is suspended vertically with its lower edge in a second solvent such as an alcohol (Figure 1.11, overleaf), then the sample (the solute) has a 'choice' of dissolving in the water which is part of the paper's structure or in the alcohol as it ascends by capillary action. If the solute is an ink, which typically contains several dyes, the dyes will then separate, and coloured traces will be seen on the paper.

Figure 1.11
*Paper
chromatography.*

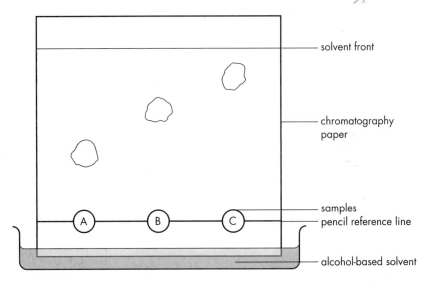

As the solvent front moves up the paper, it passes over the dots of the samples. The sample components which are most soluble in the alcohol will travel the furthest up the paper (e.g. sample C in Figure 1.11). The distance the component has travelled can be compared with the distance the solvent front has gone. The resulting ratio is called the R_f factor. It is possible to consult data and use the measured R_f value to verify the nature of the sample tested. This technique can be used to identify, say, the colouring matter in foodstuffs or the presence of certain drugs in blood or urine. Another example is its use to identify the presence of ketones in the urine of children suffering from phenylketonuria.

Chromatography is a straightforward technique provided it is done with care and cleanliness. Good 'house rules' are:

- avoid touching the face of any chromatography paper to prevent contamination
- handle the paper by its corners or, if small scale, between the fingers as it if were a photographic negative
- place a piece of clean scrap paper under the paper to prevent contamination from the bench
- use a pencil to draw the reference line on which samples will be placed – pencil is unaffected by any solvent and so will not interfere with the chromatogram
- avoid the use of ballpoints or individual single coloured felt pens – the former are often impervious to solvents and the latter often contain only one dye
- keep the developing solvents in a fume cupboard – they are often unpleasantly smelly.

Investigation of felt-pen inks

Materials
- eye protection
- square chromatography paper with punched holes
- chromatography frame, tray and tank
- ruler
- pencil
- clean scrap paper
- selection of felt pens, including black
- access to fume cupboard
- solvent made from
  3 volumes of butan-1-ol
  1 volume of ethanol (industrial methylated spirit)
  1 volume of 2 mol/dm^3 ammonia solution
  (Hint: use 200 cm^3 as '1 volume'; mix and store the total mixture in a 'Winchester' quart bottle.)

Safety
- *Butan-1-ol is highly flammable and is harmful by inhalation.*
- *Ethanol is highly flammable. Industrial methylated spirit contains methanol, which is toxic.*
- *The solvent mixture is unpleasantly smelly. Keep the stoppered bottle in a fume cupboard.*
- *The chromatograms could be run in the open laboratory if in a closed container (for example, a stoppered test-tube), but would be better carried out in a fume cupboard.*
- *Wear eye protection when pouring out the solvent.*

Figure 1.12
Paper chromatography of inks.

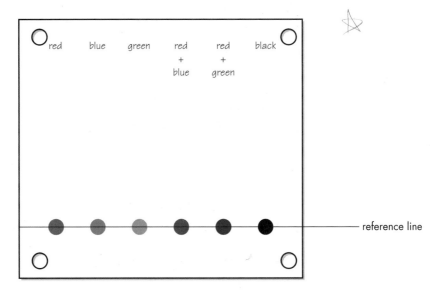

Procedure

1. Place the chromatography paper on the scrap paper on a flat surface.
2. Draw a *pencil* reference line across the paper, 2 cm from the base and parallel with it (see Figure 1.12).
3. In pencil, mark six dots on the line, at least 2 cm apart.
4. Put a different ink sample on each dot, building up the dot to a diameter no more than 0.5 cm. (It is best to let a dot dry for a short time and then add to it – too wide a drop will interfere with the adjacent one as it moves up the paper.)
5. Write the colours at the *top* of the paper opposite to the dots for reference.
6. Load the sheet on to the chromatography frame with a spacer at each corner, and with all the colours on the same edge.
7. Pour about 1 cm depth of solvent into the tray inside the tank in the fume cupboard.
8. Stand the frame in the solvent so that the papers are submerged to a depth of about 1 cm.
9. Put the cover on the tank and leave it for about 30–40 minutes until the solvent has almost reached the top of the paper.
10. Lift out the tray and leave it to dry.

Simple individual chromatogram

An alternative simple test-tube experiment is to use a drop of black ink (Quink works well) on a strip of chromatography paper. This is pinned with a drawing pin to the base of a cork which fits a boiling tube (Figure 1.13).

Figure 1.13
Paper chromatography in a test-tube.

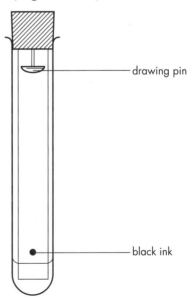

drawing pin

black ink

Procedure

1. Add enough solvent (see page 21 for formulation of solvent) to fill the curved base of the tube. Make sure the solvent does not dribble down the sides of the tube, that the paper does not touch the sides of the tube and that the tube is not shaken.

2. Put the tube in a stand while the solvent rises (about 30 minutes).

3. Take the cork out when the solvent is near the top and allow the paper to dry in a fume cupboard.

Sublimation

This term is used to describe the change of state from solid to gas without going through a liquid state. A simple everyday example is a piece of hard cheese that gives off a smell, implying that particles are escaping from the solid as a gas or vapour. Good laboratory examples are ammonium chloride or iodine, which sublime easily.

Sublimation of ammonium chloride or iodine

Materials
- eye protection
- Pyrex test-tube or ignition tube
- Bunsen burner
- retort stand (optional) or test-tube holder
- ammonium chloride
- mineral wool
- (tweezers)
- (iodine crystals)

Safety
- *Ammonium chloride is harmful and irritating to eyes.*
- *Iodine is harmful by inhalation and skin contact and causes burns over time.*
- *Iodine vapour crystallises very painfully on eyeballs.*
- *Use heat-resistant (borosilicate) glass not soft (soda) glass tubes which might distort in the flame.*
- *If the test-tube is hand-held, use a holder (which could be made from folded paper) and point it away from other people.*
- *Use mineral wool to plug the test-tube.*
- *Wear eye protection.*

Figure 1.14
Investigating the sublimation of ammonium chloride.

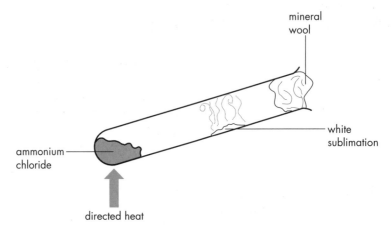

mineral wool

white sublimation

ammonium chloride

directed heat

<u>Procedure</u>
1. Put enough ammonium chloride into a test-tube to just fill the base. Plug the tube with mineral wool.
2. Warm it in a small (blue) Bunsen flame so that the heat is concentrated at the base of the test-tube, to keep the open end of the test-tube relatively cold (see Figure 1.14). (Use of a retort stand to hold the test tube allows the heat to be more accurately directed.)

<u>What you might expect</u>
White fumes of ammonium chloride will appear in the test-tube and settle out in the colder part as a white powder.

Alternatively, use tweezers to place two or three crystals of iodine in an ignition tube. These are silvery-grey when looked at individually, but they have a purplish hue in bulk. Use the same technique as above. A purple vapour is quickly seen which solidifies again as sparkly silver crystals at the top of the tube. It is safer to do this with the ignition tube clamped in a retort stand to avoid burnt fingers. The mineral wool plug is essential to prevent the purple vapour escaping.

♦ *Further activities*

- ♦ Demonstration of gas chromatography (No. 24 in *Classic Chemistry Demonstrations*, Royal Society of Chemistry, 1998, ISBN 1 870343 38 7).
- ♦ Melting points can be used to test the purity of compounds: a pure compound has a distinct melting point, but an impure substance melts at a lower temperature and over a range of temperatures. (See No. 40, Observing the lowering of a melting point, in *Microscale Chemistry*, Royal Society of Chemistry, 1998, ISBN 1 870343 49 2).

1.3 Hardness in water

♦ *Previous knowledge and experience*

Pupils may be familiar with the furring-up of kettles and the formation of scum when using soap in hard water areas. Some may have water softeners at home. They may have noticed a difference in water quality when on holiday in different areas.

♦ *A teaching sequence*

Water is such a good solvent that it is difficult to obtain it 'pure'. Rivers and streams have long since dissolved the most soluble rocks, such as sodium chloride, but running water can still attack less soluble ones and dissolve enough of them to cause inconvenience and expense when using it.

Soap is a soluble sodium salt of an organic acid such as stearic acid. If there are dissolved calcium or magnesium ions in the water, however, these will react with the soap and form an insoluble scum or curd, preventing the soap from doing its proper 'cleaning' job. Any water which contains sufficient calcium or magnesium ions to cause this scum is known as 'hard' water.

Hard water does not interfere with soapless detergents. Soapless detergents are produced by the action of concentrated sulphuric acid on hydrocarbons. Both soaps and soapless detergents contain molecules with a long hydrocarbon chain with an ionic group attached to one end. In a soap the ionic group is a carboxylate group and in a soapless detergent it is a sulphonate group.

If water percolates through gypsum ($CaSO_4.2H_2O$), it dissolves some of it and becomes a permanently hard water (PHW), so named because it cannot lose this hardness on boiling.

If rain falls on a chalky or limestone area, both of which are mainly calcium carbonate, then the water with its dissolved carbon dioxide can react over the years with the carbonate rock to form a soluble salt called calcium hydrogencarbonate:

$$CaCO_3 + H_2O + CO_2 \rightarrow Ca(HCO_3)_2$$

If water containing calcium hydrogencarbonate is boiled, the calcium hydrogencarbonate is converted back into insoluble calcium carbonate – the white deposit found inside kettles, called 'kettle fur'. Water with this variety of hardness is known as temporarily hard water (THW).

Water which does not contain hardness is known as 'soft water'.

Hardness in water is

- wasteful of soap
- wasteful of energy if it forms insulating deposits inside pipes
- the cause of blockage in water pipes in older houses.

It is beneficial in that it is

- a source of calcium and magnesium in the diet (used for growing bones and teeth)
- said to improve the taste of beer.

Hardness in water

Materials
- eye protection
- test-tube fitted with delivery tube in a rubber bung
- test-tube to hold lime water
- test-tube holder
- Bunsen burner
- calcium hydroxide solution (lime water)
- small marble chip (calcium carbonate)
- dilute hydrochloric acid (0.4 mol/dm^3)

Safety
- *Warming the final solution can cause ejection from the tube if it is boiled too vigorously or if the tube is more than one-fifth full. Make sure the tube is not pointing at anyone.*
- *Eye protection must be worn. Acid on carbonates can cause a spray and blowing through lime water can cause the lime water to be blown from the test-tube.*

Procedure
1. Put about 4 cm depth of lime water into a test-tube.
2. Put the marble chip into the second test-tube and add enough dilute acid to half-fill the test-tube.
3. Immediately fit the delivery tube and put its end *just* under the surface of the lime water (see Figure 1.15). If put in too deeply, back-pressure can sometimes stop the gas bubbling through. Make sure that the test-tube with the acid in is not allowed to be horizontal, or the acid will get into the lime water.

Figure 1.15
Bubbling carbon dioxide through lime water to produce a solution of calcium hydrogencarbonate.

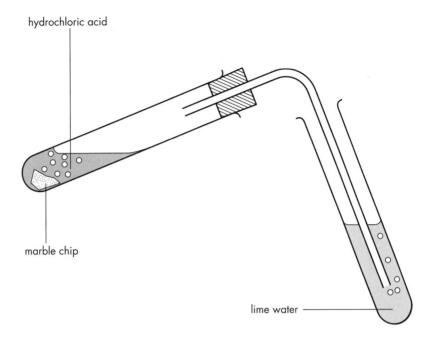

hydrochloric acid

marble chip

lime water

4. Remove the delivery tube from the lime water. Cautiously, warm it in a gentle blue Bunsen flame (air-hole half open) by moving the test-tube in and out of the side of the flame: 15–20 seconds should be long enough. The test-tube should not be more than one-fifth full to reduce risk of spurting out. Putting the test-tube in the middle of the flame and heating it strongly will produce too vigorous a reaction and will eject the contents of the test-tube.

<u>What you might expect</u>
In step 3, bubbles of carbon dioxide will pass through the lime water and turn it cloudy, because insoluble calcium carbonate is being made. If the carbon dioxide is allowed to continue bubbling through, the cloudiness disappears; this is because the calcium carbonate is being converted into soluble calcium hydrogencarbonate. The disappearance of the cloudiness illustrates the formation of temporary hardness in water. In step 4, the cloudiness should start to reappear as the calcium hydrogencarbonate turns back into insoluble calcium carbonate. This reappearance of the cloudiness is the same reaction as the removal by boiling of temporary hardness in water, accompanied by the appearance of 'kettle fur'.

An alternative to using acid on a marble chip is to allow pupils to exhale air through a straw into a test-tube of lime water. They should be warned not to blow too hard or they will blow the lime water out of the tube in their face; pupils should be advised to wear eye protection.

The sequence

calcium carbonate + water + carbon dioxide $\rightleftharpoons$ calcium hydrogencarbonate

is also responsible for the formation of stalactites and stalagmites. The drips of calcium hydrogencarbonate in a limestone cave decompose over the years to produce a growth of insoluble calcium carbonate.

Comparison of hardness in water samples

<u>Materials</u>
- burette in a burette stand
- 100 cm^3 conical flask
- 10 cm^3 pipette and pipette filler (or 10 cm^3 measuring cylinder)
- soap solution
- sample A – distilled water
- sample B – distilled water after shaking with calcium sulphate and filtering (to simulate PHW)
- sample C – source of THW
- sample D – source of THW which has been boiled until the hardness is softened, and then filtered

<u>Safety</u>
- *Many soap solutions are made up in ethanol, which is highly flammable.*
- *In the interest of hygiene, use a pipette filler with the pipette. If one is not available, use a measuring cylinder with the filling achieved accurately by using a pipette.*

In this experiment, the amount of hardness present is assessed by measuring the volume of soap solution used to produce a permanent lather.

<u>Procedure</u>
1. Measure out 10 cm^3 of one of the samples of water into the conical flask. It is important to use the same volume for each sample of water, for this to be a 'fair test'.
2. Add 1 cm^3 of the soap solution from the burette. Shake the conical flask thoroughly.
3. Continue adding soap solution 1 cm^3 at a time with shaking until a 'permanent lather' is obtained. This is defined (arbitrarily) as a continuous lather over the surface of the liquid which lasts for about 20 seconds.

<u>What you might expect</u>

The amount of soap solution used with distilled water (sample A) is simply the quantity needed to make a lather; it is likely to be about 1 cm³. The other amounts, less the amount needed to make the lather with distilled water, will be a measure of the hardness present:

sample B − sample A = PHW present in this artificial source
sample C − sample A = THW present
sample D − sample C = any PHW which may be present

(If sample D has the same value as sample A, it means there is no PHW in the sample.)

(*Note*: if it is known that the local tap water contains both PHW and THW, it is not necessary to use sample B in the experiment.)

Methods of softening water

- Boiling: this removes THW only.
- Adding sodium carbonate: this removes THW and PHW. The carbonate ions react with the calcium (or magnesium) ions and precipitate out the hardness as insoluble calcium (or magnesium) carbonate.
- Using an 'ion exchange' column: this is the basis of household water softeners. The water is allowed to percolate downwards through a zeolite resin in a column. The Ca^{2+} and Mg^{2+} ions are exchanged for H^+ ions in the resin. Typically, the resin is regenerated by standing it overnight in brine solution, which replaces the Ca^{2+} and Mg^{2+} ions by Na^+ ions. The first water through the softener in the morning has to be discarded to dispose of the unwanted ions. Other resins can remove SO_4^{2-} and HCO_3^- ions, producing 'deionised' water, which is the 'purest' available.

◆ *Enhancement ideas*

Water hardness tests

Hardness of water from different sources can be compared using soap solution as above. Alternatively, test strips can be purchased which will give a reliable measure of total water hardness in 15 seconds. Total hardness is measured in a range 0–450 ppm. These strips are sold in packs of 250 or 1000 and are ideal for fieldwork. They are easily obtained from the Mid-City Supply Company, Main Street, Elkhart, Indiana, USA or **sales@mid-city.com**; further information can be obtained from **www.mid-city.com**.

1.4 Elements and compounds

◆ *Previous knowledge and experience*

Pupils may have performed some chemical reactions and used the chemical names of some elements and compounds but they are unlikely to be familiar with the abstract ideas of elements and compounds.

◆ *A teaching sequence*

Pupils are often confused by terminology in chemistry. It is important that you define some key words accurately and clearly, and cultivate their proper use.

An *element* is a substance which cannot be decomposed in a chemical reaction into two or more simpler substances.

All elements are built up from *atoms*. An atom is the smallest particle of an element which can take part in a chemical reaction.

Each atom of a particular element will react in the same way as any other atom of that element. The way in which an element reacts depends on the electrons it has available, especially in its outermost shell. The structure of atoms and the bonding which is used when they react together is dealt with more fully in Chapter 2.

You need to make it clear that mixing elements together does not necessarily produce a chemical reaction: sometimes it just produces a *mixture*. For example, air is a mixture of gases, containing predominantly nitrogen (about 78%), oxygen (about 20%), argon (about 1%), carbon dioxide (about 0.03%) and traces of other noble gases.

A mixture

- can be separated into its components by suitable physical means
- will have a variable composition, depending on how much of the different components are mixed together
- has a chemical behaviour that depends on its components.

As an example, a mixture of iron and sulphur can be separated with a magnet. It has the same properties as iron and sulphur.

When elements do combine together, they produce *compounds.*

A compound

- is a substance which contains two or more elements chemically combined together
- can only be split into its component elements by chemical means
- has a definite chemical composition and thus a specific formula
- has its own chemical behaviour, which is different from the chemical behaviour of any of the elements present in it.

Iron sulphide is the compound formed from iron and sulphur.

Distinguishing between a mixture and a compound

Materials

- eye protection
- 3 soft glass test-tubes
- small bar magnet
- paper tissue
- watch glass (5 cm diameter)
- pestle and mortar
- Bunsen burner
- mineral wool
- matches
- test-tube holder
- test-tube rack
- splints
- iron/sulphur mixture (7 : 4 ratio by mass)
- hydrochloric acid (0.4 mol/dm^3)
- water

Safety
- *Sulphur is a fine powder which irritates the eyes.*
- *Hydrogen sulphide is toxic (nearly as toxic as hydrogen cyanide), but is easily recognised by its 'bad egg' smell. Use a fume cupboard when testing for hydrogen sulphide.*
- *Sulphur vapour is flammable, forming sulphur dioxide gas. Sulphur dioxide is toxic and corrosive; it may also trigger an asthma attack in a sensitive individual. Use a fume cupboard when heating the iron/sulphur mixture.*
- *Wear eye protection when using the iron/sulphur mixture.*

Procedure

1. Iron and sulphur can be separated by physical means. Wrap the end of a magnet in a paper tissue and put it into a teaspoon-sized heap of the mixture on a watch glass. The iron is attracted but the sulphur stays behind.

2. Put about 1 cm depth of the mixture into a test-tube and half-fill it with tap water. Shake it vigorously for about 10 seconds, then allow it to stand in the rack. The less dense sulphur tends to float, and the denser iron sinks to the bottom.

3. Put about 1 cm depth of mixture into a test-tube and half-fill the test-tube with dilute hydrochloric acid. After a while hydrogen gas is evolved, which can be tested with a lighted splint (the 'squeaky pop' test).

    Hint: Keep the open end of the test-tube covered by your thumb until you feel the gas pressure before trying to test for hydrogen with a lighted splint. Otherwise the gas just escapes and no pop will be heard.

Demonstration

You can then demonstrate what happens when a test-tube of the mixture is heated. It helps if the mixture is slightly rich in sulphur, because some tends to vaporise out instead of reacting with the iron.

Figure 1.16

Heating a mixture of iron and sulphur.

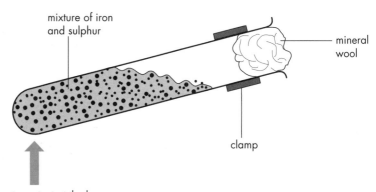

mixture of iron and sulphur

mineral wool

clamp

keep heat constant at the base

1. Half-fill a soft glass test-tube with the iron–sulphur mixture. Hold it in a test-tube holder near its neck or clamp it in a stand. Plug the mouth of the test-tube with mineral wool.

2. Heat it gently, keeping the flame directed constantly at the base of the test-tube (see Figure 1.16), until an orange glow is seen inside the tube. Immediately stop the heating and let the pupils see that the glow continues and steadily rises through the test-tube.

3. Allow the contents to cool and then break the test-tube with a pestle inside the mortar.

Naming of compounds

You need to emphasise to pupils that elemental gases such as hydrogen, oxygen and nitrogen contain two atoms combined together, but they are not usually referred to as compounds because only one element is involved in each.

If only two elements are combined together, the name ends in '-ide'. It is customary to name any metal first; for example, sodium and oxygen produce sodium oxide and magnesium and bromine produce magnesium bromide. Pupils may also come across oxides of non-metals such as carbon, nitrogen and sulphur.

If three elements are combined in a compound and one is oxygen, the name ends in '-ate'. Pupils will meet the following: metal nitrates (metal + nitrogen + oxygen), metal sulphates (metal + sulphur + oxygen) and metal carbonates (metal + carbon + oxygen).

An –OH group (containing oxygen and hydrogen) is called a hydroxyl group but the compound in which it is found is a hydroxide, e.g. KOH is potassium hydroxide.

The formula Na_2CO_3 represents one 'formula unit' of sodium carbonate, containing two atoms of sodium, one atom of carbon and three atoms of oxygen, i.e. the number of atoms of the element is shown as a subscript immediately after its symbol.

The reason why elements combine in certain proportions is explained in Chapter 2, *Bonding* (page 66). The formulae for the compounds between elements in one group and elements in another group follow a pattern. Table 1.4 gives a 'template' for some common substances. It is not an exhaustive list but sufficient for work at this level.

Table 1.4 *Examples of formulae of compounds.*

Group 1		Group 2		Group 3	
Sodium oxide	Na_2O	Calcium oxide	CaO	Aluminium oxide	Al_2O_3
Sodium hydroxide	$NaOH$	Calcium hydroxide	$Ca(OH)_2$	Aluminium hydroxide	$Al(OH)_3$
Sodium chloride	$NaCl$	Calcium chloride	$CaCl_2$	Aluminium chloride	$AlCl_3$
Sodium nitrate	$NaNO_3$	Calcium nitrate	$Ca(NO_3)_2$	Aluminium nitrate	$Al(NO_3)_3$
Sodium sulphate	Na_2SO_4	Calcium sulphate	$CaSO_4$	Aluminium sulphate	$Al_2(SO_4)_3$
Sodium carbonate	Na_2CO_3	Calcium carbonate	$CaCO_3$	Aluminium carbonate	$Al_2(CO_3)_3$

The formulae for lithium and potassium compounds can be written down by substituting Li or K for Na, and those for magnesium compounds by substituting Mg for Ca, etc.

All transition elements have compounds with formulae like those of the Group 2 metals (see Chapter 7 about the Periodic table). This is because they have two outermost electrons which can be used in bonding. They can sometimes use other electrons too; for example, iron forms two well-defined series of compounds:

- iron(II) compounds such as FeO, $FeCl_2$ and $FeSO_4$
- iron(III) compounds such as Fe_2O_3, $FeCl_3$ and $Fe_2(SO_4)_3$.

The Roman numerals in parentheses actually refer to the number of electrons used by iron in bonding in each of these series. Iron(II) and iron(III) compounds used to be called ferrous and ferric compounds, respectively.

Writing correct chemical formulae is an important skill which requires practice. Having given the pupils a template using sodium, calcium and aluminium compounds, they can practise writing chemical formulae by filling in Table 1.5.

Table 1.5 *Practise in writing formulae.*

	Potassium, K^+	Magnesium, Mg^{2+}	Chromium, Cr^{3+}
Oxide, O^{2-}			
Hydroxide, OH^-			
Chloride, Cl^-			
Nitrate, NO_3^-			
Sulphate, SO_4^{2-}			
Carbonate, CO_3^{2-}			

At this point further practice can be obtained from simple games. For example, divide pupils into teams and give each team a series of cards with chemical formulae written on them, some of which are correct and some incorrect. Each team has to work out the name of the compounds and decide which formulae are correct and which are incorrect.

There is a tendency to leave learning to write chemical formulae until a very late stage. If it is taught early on, it is easy to make it fun and gives plenty of time during the course for pupils to practise writing formulae whenever chemicals are used.

◆ *Enhancement ideas*

◆ An exhibition of a wide range of elements and compounds could be set up. Pupils could separate the elements from the compounds and, using the names, work out which elements are present in the compounds. They will notice that sometimes two compounds can be found that contain the same elements but are very different in their properties, e.g. sugar and ethanol. You could also include some mixtures in the exhibition.

◆ Pupils could research lists of elements produced by chemists in previous centuries, for example John Dalton's list of elements and symbols (see Figure 1.17). They will find that some substances, which were once regarded as elements because they could not be split up, are now known to be compounds; examples include lime and strontian.

Figure 1.17
John Dalton's list of the elements, 1808.

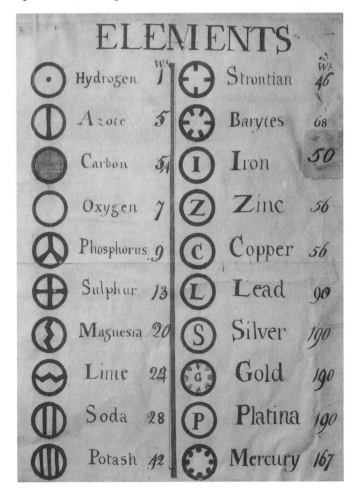

1.5 Acids, bases and salts

◆ *Previous knowledge and experience*

Pupils are unlikely to have come across this classification of compounds.

◆ *A teaching sequence*

The 'old' definition of an acid is a substance which contains hydrogen that may be replaced by a metal to produce a salt; for example, dilute hydrochloric acid produces hydrogen and the salt magnesium chloride when it reacts with magnesium.

$$Mg + 2HCl \rightarrow MgCl_2 + H_2$$

In practice the modern definition is that an acid is a proton provider; hydrochloric acid ionises in water to produce hydrogen ions and chloride ions.

$$HCl(aq) \rightarrow H^+(aq) + Cl^-(aq) \tag{1}$$

In the case of water itself, the ionisation (which is incomplete) provides H^+ and OH^- ions in equal quantities, and so water is a neutral substance.

$$H_2O(l) \rightleftharpoons H^+(aq) + OH^-(aq) \tag{2}$$

The opposite of an acid is a base, that is a metallic oxide or hydroxide. If the base is soluble in water it is called an alkali, e.g. NaOH.

$$NaOH(aq) \rightarrow Na^+(aq) + OH^-(aq) \tag{3}$$

Hence the modern definition of a soluble base is that it is a proton acceptor or OH^- provider.

Combining equations (1) and (3) from above:

$$HCl(aq) + NaOH(aq) \rightarrow NaCl(aq) + H_2O(l)$$

This can be written in an ionic form as:

$$H^+(aq) + Cl^-(aq) + Na^+(aq) + OH^-(aq) \rightarrow Na^+(aq) + Cl^-(aq) + H_2O(l)$$

Some ions, called spectator ions, occur on both sides of the equation. An equation should summarise change, so these ions can be removed from both sides of the equation.

The simplest ionic equation is

$$H^+(aq) + OH^-(aq) \rightleftharpoons H_2O(l)$$

i.e. acid + base reactions are examples of neutralisation.

Two useful summarising diagrams are shown in Figures 1.18 and 1.19 (overleaf).

Figure 1.18
Acids and bases –
a summary.

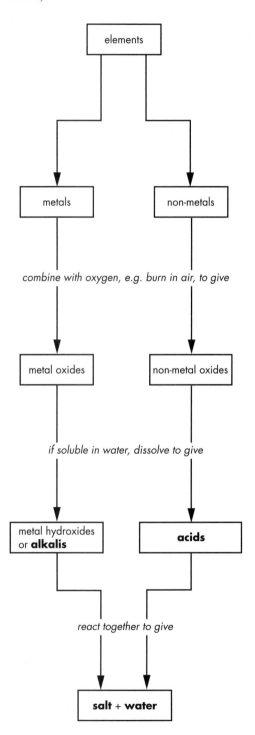

Figure 1.19
*Bases – a
summary.*

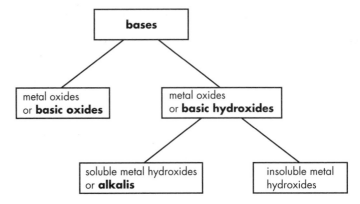

Acidity is measured on the pH scale; the pH is the logarithm to the base 10 of the hydrogen-ion concentration in moles per litre, with the sign changed. Mathematically, $pH = -\log_{10}[H^+]$.
It can be shown that in water there are 10^{-7} moles per litre of both H^+ ions and OH^- ions; hence a neutral solution has a pH value of 7. A strong acid (one which is highly ionised) has a pH of 1 and a strong alkali a pH of 13:

pH scale

pH value: 1 2 3 4 5 6 7 8 9 10 11 12 13

strong   neutral   strong
acid   alkali

A common misconception is to confuse the terms 'weak' and 'dilute' and the terms 'strong' and 'concentrated'.

*   A *weak* acid is one which is not very ionised.
*   A *dilute* acid is one with not much acid dissolved in the water.

*   A *strong* acid is one which is highly ionised.
*   A *concentrated* acid is one with a large amount of acid dissolved in the water.

Neutralisation

There are three general reactions with acids which often occur in Chemistry up to 16+:

acid + metal → salt + hydrogen
acid + base → salt + water
acid + metal carbonate → salt + water + carbon dioxide

These reactions may be used to prepare salts; full practical details may be found in standard texts. The principle is to add sufficient of the metal, base or metal carbonate to the acid until the latter is neutralised. This is usually ensured by adding excess of the solid to the acid and filtering off the excess.

The sequence is shown in Figure 1.20.

Figure 1.20
Preparing a salt.

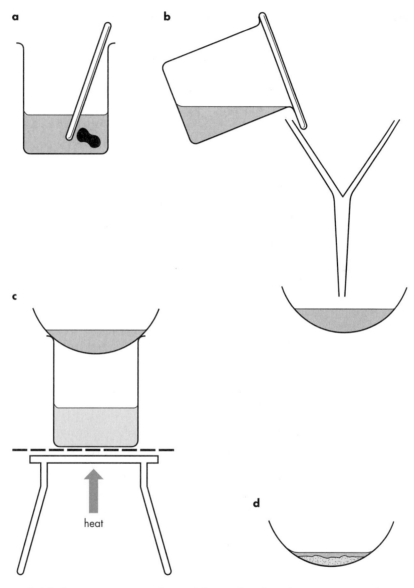

1. Add the reagent to the acid in a beaker (Figure 1.20a).
    If using a metal or a carbonate, stir until no more gas
    bubbles are produced; if using a solid base, stir with gentle
    warming, until no more will dissolve.
2. Filter to remove the excess reagent (as the residue left in
    the filter paper, Figure 1.20b).
3. Warm on a water bath to concentrate the filtrate (the salt
    in solution, Figure 1.20c).
4. Leave to crystallise (Figure 1.20d).

If the base is an alkali then a *titration* method is used instead (see Chapter 5). An example is the titration of hydrochloric acid from a burette against sodium hydroxide solution measured by a pipette, with an indicator such as methyl orange.

If the salt is insoluble in water, then it is better to use a *precipitation* method, that is, to mix two soluble salt solutions together and filter the results. The residue will be the insoluble salt, with the filtrate containing another soluble salt. For example:

lead(II) nitrate(aq) + sodium carbonate(aq)

$\rightarrow$ sodium nitrate(aq) + lead(II) carbonate(s)

◆ *Enhancement ideas*

- ◆ Pupils could test household substances with universal indicator solution or universal indicator paper and classify them as acidic, alkaline or neutral. Suitable substances include tap water, distilled water, ammonia solution, washing powder, washing-up liquid, lemon juice, toothpaste and milk.

- ◆ The pH value of soil could be found by mixing soil with water and decanting off the solution (barium sulphate can be added to help clear the solution). The liquid could then be tested with Universal indicator paper (narrow range is best) or with a pH meter.

- ◆ Pupils could examine soil-testing kits sold in horticultural shops. They could research and find out which plants grow best in soils with different pH values.

- ◆ Indicators could be extracted from plant and vegetable material using industrial methylated spirit (IMS, colourless). Care should be taken that there are no flames as IMS is highly flammable; it can be heated by immersing the test-tube in hot water (e.g. from a kettle). (IMS contains methanol which is toxic.) Red cabbage and beetroot work well and are available all through the year. Most flowers work well unless they are yellow in colour.

1.6 Equations

♦ *Previous knowledge and experience*

Pupils may have come across some simple word equations to describe reactions such as combustion or photosynthesis.

♦ *A teaching sequence*

Word equations

Word equations should be seen by pupils as a way of summarising a chemical reaction, with the reacting substances (reactants) on the left-hand side and the substances produced (products) on the right-hand side. An arrow is used rather than an equals sign because the two sides are not strictly equal and an arrow shows the direction of the reaction. Pupils should be encouraged to write word equations whenever they come across a chemical reaction. For example:

magnesium oxide + sulphuric acid → magnesium sulphate + hydrogen

For many pupils, writing word equations will be the limit of their equation writing but for pupils who go on to write symbol equations, the word equation is still an important precursor. When pupils start to write equations for reactions (i.e. the 'recipe' by which a reaction works) they must be taught *not* to alter the accepted formulae. In a chemical reaction, sodium hydroxide will *always* have the formula $NaOH$, calcium nitrate will *always* be $Ca(NO_3)_2$, and so on.

Balanced symbol equations

More able pupils, and certainly pupils going on to study Chemistry at a higher level, will need to be able to write balanced symbol equations. Even students who are not able to write balanced symbol equations should be able to extract information from an equation. A balanced symbol equation uses the correct formula for each substance and makes sure that every atom is accounted for, i.e. that there are the same number of atoms of a particular element on both sides of the equation. This is achieved by altering the ratio of molecules in the equation as necessary.

Remember that the elemental gases hydrogen, oxygen, nitrogen and halogens contain two atoms in their molecules, i.e. H_2, O_2, N_2, F_2 and Cl_2.

Consider the reaction between zinc and hydrochloric acid:

zinc + hydrochloric acid → zinc chloride + hydrogen

Substituting formulae for the names gives

Zn	+	HCl	→	ZnCl$_2$	+	H$_2$
1 atom of zinc		1 molecule of hydrochloric acid, containing 1 atom of hydrogen and 1 atom of chlorine		1 molecule of zinc chloride, containing 1 atom of zinc and 2 atoms of chlorine		1 molecule of hydrogen, containing 2 atoms of hydrogen

Now count the atoms on both sides.

Reactants: 1 atom of zinc   Products: 1 atom of zinc
1 atom of hydrogen   2 atoms of chlorine
1 atom of chlorine   2 atoms of hydrogen

But for the equation to balance, there must be the same number of atoms of a particular element on both sides of the equation. This can be achieved by putting *two* molecules of hydrochloric acid into the equation:

Zn + 2HCl → ZnCl$_2$ + H$_2$ ✓

Note: The use of brackets around a group in a formula is a convention used to show how many of that group have to be present. ZnCl$_2$ does not need brackets because Cl represents just one element. But calcium hydroxide is written as Ca(OH)$_2$ (pronounced 'C-A, OH twice') because two –OH groups are needed.
   The frequent mistake pupils make when balancing equations is to alter the formulae, for example:

Zn + HCl → ZnCl + H ✗

Balancing equations needs practice. Introducing symbol equations as soon as possible enables pupils to practise equation writing and balancing through the rest of the course.

The following list of equations summarises many of the common reactions found in secondary schools and can be used as the basis of equation-writing practice.

$Fe(s) + S(s) \rightarrow FeS(s)$

$2Al(s) + 3I_2(s) \rightarrow 2AlI_3(s)$

$2H_2(g) + O_2(g) \rightarrow 2H_2O(l)$

$2H_2O_2(aq) \rightarrow 2H_2O(l) + O_2(g)$

$2Cu(s) + O_2(g) \rightarrow 2CuO(s)$

$CaCO_3(s) + H_2O(l) + CO_2(g) \rightarrow Ca(HCO_3)_2(aq)$

$Mg(s) + H_2O(g) \rightarrow MgO(s) + H_2(g)$

$Zn(s) + 2HCl(aq) \rightarrow ZnCl_2(aq) + H_2(g)$

$Zn(s) + H_2SO_4(aq) \rightarrow ZnSO_4(aq) + H_2(g)$

$PbO(s) + H_2(g) \rightarrow Pb(s) + H_2O(g)$

$C(s) + O_2(g) \rightarrow CO_2(g)$

$CO_2(g) + C(s) \rightarrow 2CO(g)$

$Fe_2O_3(s) + 3CO(g) \rightarrow 2Fe(l) + 3CO_2(g)$

$CaCO_3(s) \rightarrow CaO(s) + CO_2(g)$

$CaO(s) + SiO_2(s) \rightarrow CaSiO_3(l)$

$Fe_2O_3(l) + 2Al(s) \rightarrow 2Fe(s) + Al_2O_3(s)$

$CuSO_4(aq) + Fe(s) \rightarrow Cu(s) + FeSO_4(aq)$

$2Na(s) + 2H_2O(l) \rightarrow 2NaOH(aq) + H_2(g)$

$2K(s) + 2H_2O(l) \rightarrow 2KOH(aq) + H_2(g)$

$2Na(s) + Cl_2(g) \rightarrow 2NaCl(s)$

$H_2(g) + Br_2(g) \rightarrow 2HBr(g)$

$Cl_2(g) + 2KBr(aq) \rightarrow Br_2(aq) + 2KCl(aq)$

$(NH_4)_2SO_4(s) + 2NaOH(aq) \rightarrow 2NH_3(g) + Na_2SO_4(aq) + 2H_2O(l)$

$N_2(g) + 3H_2(g) \rightleftharpoons 2NH_3(g)$

$2NH_3(g) + H_2SO_4(aq) \rightarrow (NH_4)_2SO_4(aq)$

$4NH_3(g) + 5O_2(g) \rightarrow 4NO(g) + 6H_2O(g)$

$2NO(g) + O_2(g) \rightarrow 2NO_2(g)$

$C_2H_5OH(g) \rightarrow H_2O(g) + C_2H_4(g)$

$H_2O(l) + C_{12}H_{22}O_{11}(aq) \rightarrow 4C_2H_5OH(aq) + 4CO_2(g)$

$CuO(s) + H_2SO_4(aq) \rightarrow CuSO_4(aq) + H_2O(l)$

$CaCO_3(s) + 2HCl(aq) \rightarrow CaCl_2(aq) + CO_2(g) + H_2O(l)$

$Mg(s) + H_2SO_4(aq) \rightarrow MgSO_4(aq) + H_2(g)$

$Ba(OH)_2(aq) + H_2SO_4(aq) \rightarrow BaSO_4(s) + 2H_2O(l)$

$Na_2S_2O_3(aq) + 2HCl(aq) \rightarrow 2NaCl(aq) + SO_2(g) + S(s) + H_2O(l)$

$3Fe(s) + 4H_2O(g) \rightleftharpoons Fe_3O_4(s) + 4H_2(g)$

$CH_4(g) + 2O_2(g) \rightarrow CO_2(g) + 2H_2O(l)$

$2CH_4(g) + 3O_2(g) \rightarrow 2CO(g) + 4H_2O(l)$

State symbols

Pupils may be asked to include state symbols in the equations they write. These are:

- solid (s)
- liquid (l)
- gas (g)
- solution in water (aq) (from the Latin word *'aqua'*, meaning water).

Remember that if a precipitate appears its state symbol will be (s).

Hence, if zinc metal is added to hydrochloric acid, the equation including state symbols is:

$$Zn(s) + 2HCl(aq) \rightarrow ZnCl_2(aq) + H_2(g)$$

The difference between water and steam is shown by $H_2O(l)$ and $H_2O(g)$. Take care that pupils do not confuse a solution with 'being a liquid' and therefore try to use (l) out of context: the symbol (l) is reserved for substances which are liquids in their own right, e.g. mercury, $Hg(l)$, and ethanol, $C_2H_5OH(l)$.

As a general rule, state symbols are not required in examinations unless they are specifically asked for in the question.

Ionic equations

The most able pupils are required to learn how to write ionic equations. They follow the same rules for balancing as above but, in addition, they must be balanced in terms of charge. They should omit any ions which do not actually play a part in the reaction. Pupils need to realise that:

- if an ionic substance dissolves in water it becomes split up into its component ions, i.e. it becomes ionised
- solid, liquids and gases remain un-ionised in equations
- acids, alkalis and all soluble salts can be written as component ions.

As an example, if copper(II) sulphate solution and sodium hydroxide solution are mixed together, a blue precipitate of copper(II) hydroxide forms. The other product is soluble sodium sulphate. The equations are:

copper(II) sulphate	+	sodium hydroxide	$\rightarrow$	copper(II) hydroxide	+	sodium sulphate
$CuSO_4(aq)$	+	$2NaOH(aq)$	$\rightarrow$	$Cu(OH)_2(s)$	+	$Na_2SO_4(aq)$

or

$$Cu^{2+}(aq) + SO_4^{2-}(aq) + 2Na^+(aq) + 2OH^-(aq) \rightarrow$$

$$Cu(OH)_2(s) + 2Na^+(aq) + SO_4^{2-}(aq)$$

Taking out the ions which are identical on both sides of the equation gives the ionic equation:

$$Cu^{2+}(aq) + 2OH^-(aq) \rightarrow Cu(OH_2(s)$$

The omitted ions are sometimes referred to as 'spectator ions'. Common ionic equations include:

$$H^+(aq) + OH^-(aq) \rightarrow H_2O(l)$$
$$O^{2-}(s) + 2H^+(aq) \rightarrow H_2O(l)$$
$$CO_3{}^{2-}(s) + 2H^+(aq) \rightarrow H_2O(l) + CO_2(g)$$
$$Ba^{2+}(aq) + SO_4{}^{2-}(aq) \rightarrow BaSO_4(s)$$
$$Ag^+(aq) + Cl^-(aq) \rightarrow AgCl(s)$$
$$NH_4{}^+(s) + OH^-(aq) \rightarrow NH_3(g) + H_2O(l)$$

In each of these equations, the sum of the charges on the left-hand side equals the sum on the right-hand side.

Able pupils should be encouraged to write ionic equations whenever suitable opportunities arise.

◆ *Other resources*

◆ Information about the properties, uses and history of the elements can be found using the CD ROM *The Elements* produced by Yorkshire International Thomson Multimedia Ltd, Kirkstall Road, Leeds LS3 1JS. This is related to the Periodic table and includes many quiz questions.

Web sites

◆ Pupils can extend their study of filtration by visiting: **chemscape.santafe.cc.fl.us/chemscape/catofp/mixpour/ filter/filter.htm**
    This gives a description of different methods of filtration, including gravity and vacuum filtration. It includes video clips and self-test questions.

◆ A concise set of notes on oil refining from an engineer can be found by visiting: **asloley.home.mindspring.com/ref00002.html**

◆ A virtual tour of a Scottish whisky distillery by the works manager can be found by visiting: **www.glenmorangie.com**

♦ Pupils can find other chromatography experiments which can be done at home by visiting:
www.exploratorium.edu/science_explorer/ black_magic.html

♦ Information about water quality can be obtained from the Water Quality Association on:
www.wqa.org

♦ Information about different types of solution and solubility of gases and solids can be found by visiting:
www.tannerm.com/aqueous/solutions/solution.htm

♦ The methods of separating mixtures in this chapter can be extended to look at more advanced methods. More able pupils will find information about advanced methods by visiting:
ull.chemistry.uakron.edu/chemsep/contents.html

♦ Information about acids and alkalis and the pH scale can be found on:
www.chem4kids.com/chem4kids/reactions/ acidbase.html

 ## Videos

Fractional distillation of crude oil is included in the video *Industrial Chemistry for Schools and Colleges* produced by the Royal Society of Chemistry (see Chapter 9). Other videos on fractional distillation are available, including 'Oil' in the Channel 4 *Science in Focus* series (1998). This video includes the Milford Haven oil disaster, which is good if you wish to extend studies to include environmental effects of oil spillages. It also includes a good animated diagrammatic section on fractional distillation.

A good video 'Elements and compounds' is also available in the Channel 4 *Science in Focus* series (1998).

2 *Particles*

Geoff Mines

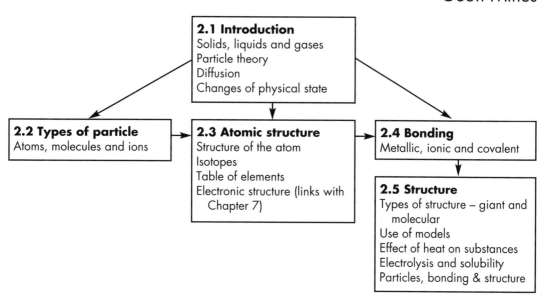

2.1 Introduction
Solids, liquids and gases
Particle theory
Diffusion
Changes of physical state

2.2 Types of particle
Atoms, molecules and ions

2.3 Atomic structure
Structure of the atom
Isotopes
Table of elements
Electronic structure (links with
    Chapter 7)

2.4 Bonding
Metallic, ionic and covalent

2.5 Structure
Types of structure – giant and
    molecular
Use of models
Effect of heat on substances
Electrolysis and solubility
Particles, bonding & structure

◆ *Choosing a route*

Scientists are continually creating new materials with novel
properties in fields such as plastics and ceramics, and these
developments depend on a detailed understanding of how the
particles are arranged and held together. This topic looks into
the structure of these particles – atoms, molecules and ions –
and the way they are bonded and arranged within substances.
    The key ideas and concepts include:

- the introduction of the particle theory to explain the properties
  of solids, liquids and gases and simple physical processes
- the different types of particle that form the building blocks
  of all materials
- the particles that make up the structure of the atom
- the bonding and structure of elements and compounds and
  their associated properties.

Many pupils can cope with the bulk properties of materials but
struggle when ideas about particles are introduced. For the
more able pupils, this introduction should start at the age of
about 12–13. Lower ability pupils may never cope with the
abstract ideas of particles.

2.1 Introduction

◆ *Previous knowledge and experience*

Pupils will have spent some time gathering information about materials at primary school. They should be able to recognise the differences in some properties between solids, liquids and gases, and a good starting point is to classify the data they have obtained.

Pupils will know that a smell spreads out to all corners of the room, and that when a coloured soluble substance is added to water the colour spreads out throughout the water. Pupils will know from primary school that substances can exist in three states, solid, liquid and gas, depending upon conditions. Water can exist as ice, liquid water and steam (or water vapour).

◆ *A teaching sequence*

Solids, liquids and gases

Pupils could fill in a blank form of a table such as Table 2.1.

Table 2.1 *Properties of the three states of matter.*

Property	Solid	Liquid	Gas
Volume of a certain quantity	Fixed	Fixed	Changes to fill the container
Shape	Fixed	Changes to fit the shape of the bottom of the container	Takes up the shape of the whole container
Density (mass of a given volume)	High	High	Low
Expansion on heating	Low	Medium	High
Ease of compression	Very low	Low	High

It is also probably a good idea at this stage to revise the physical processes involved in changing from one state to another, with triangular diagrams such as Figure 2.1.

Figure 2.1
Changes of state.

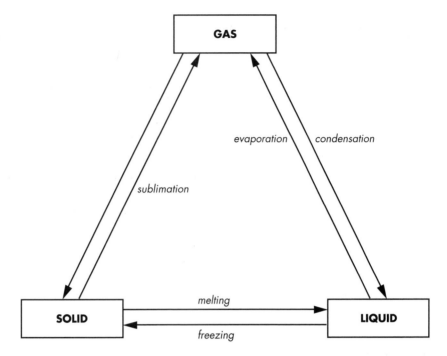

Particle theory

Having reminded pupils of the properties of solids, liquids and gases, it is time to make a significant jump in the level of demand. The key problem is now to try and explain the similarities and differences in terms of a model at the molecular level – the so-called 'particle theory'.

The key aspects of the particle theory are that all matter is made up of small, moving particles, which

- can be close together or far apart
- can be regularly or irregularly arranged
- can be held together by strong or by weak forces
- move faster as temperature increases.

 A possible approach is to describe these key aspects of the particle theory and illustrate them with suitable diagrams, models and computer simulations. Alternatively, pupils could develop their own understanding in a more individual way by identifying the 'particle pictures' in Figure 2.2 (overleaf) as solid, liquid or gas and then arrange the descriptions and properties in Table 2.2 which match in a suitable table.

At this stage it is probably best just to use the word 'particle' to cover all situations. To introduce the terms 'atoms', 'molecules' and 'ions' at the same time would possibly overload pupils with another set of terminology to cope with.

Figure 2.2
Solid, liquid or gas?

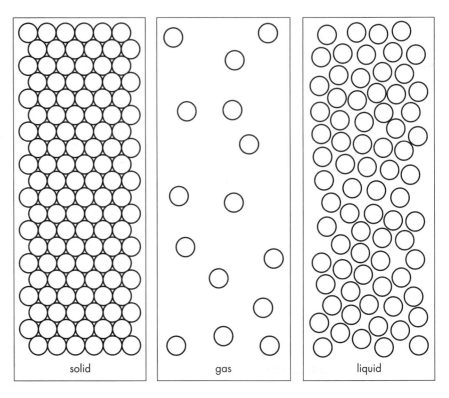

solid   gas   liquid

Table 2.2   *Solid, liquid or gas?*

Description		
Widely spread out	Regular pattern	Not held together
Weakly held together	Closely packed	Freely moving
Strongly bonded (linked) together	Vibrating on the spot	Slide over each other
Sometimes bump into each other	Not arranged regularly	
Properties		
Often strong	No fixed shape	Spread out to fill any container
Easily compressed	No strength	Little strength
Have a fixed shape	Only mix very slowly, if at all	Mix quite easily
Mix and spread out quickly	Hard to compress	

Poster work
The abstract nature of this area makes significant demands on pupils and should not be hurried since it has an important impact on future understanding of the subject. The conceptual and linguistic challenges must not be under-estimated, and time should be allowed for pupils to develop their own pictorial representations and to draw them in their own way – poster work for homework is a good idea here.

Models

The use of diagrams and models in this area is crucial to illustrate the similarities and differences between the particle arrangements and structures of solids, liquids and gases. The models can be constructed from, for example, coloured beads, 'Molymod' kits or polystyrene balls, but of course these can only give a static picture. The dynamic nature of particle theory should not be forgotten and is probably best illustrated with appropriate computer programs.

It is also a good idea at this time to explain the origin of the pressure exerted by gases in terms of particle collisions with the walls of the containing vessel. Furthermore, the effect of changes in temperature or volume on the pressure of a gas are useful illustrations of the application of particle theory.

When pupils are thoroughly familiar with the key terms and ideas, further experimental work, such as the following activities illustrating diffusion and change of state, can be carried out to provide more opportunities for them to extend and apply their knowledge.

Diffusion

Diffusion is the movement of matter to fill all of the available space. Pupils will eventually be able to explain this in terms of a particle model.

The topic of diffusion is best introduced practically, first by demonstration and then by pupil experiment. In one demonstration bromine is used, which can be seen. In the other demonstrations the diffusing matter cannot be seen but the effects of its movement can. Pupils can reinforce their understanding by experiments with carbon dioxide. Finally, they can examine diffusion in liquids, using potassium manganate(VII).

Demonstration of diffusion in air

Place some scent or perfume with a distinctive smell on a watch glass at the front of the class; ask the pupils to put up their hands when they can smell it as it spreads through the room.

Demonstration of diffusion of bromine in air

This experiment shows that bromine vapour spreads out to fill both gas jars evenly (Figure 2.3, overleaf). Before the demonstration pupils may believe that the bromine vapour, being denser than air, will continue to fill the bottom gas jar, with air filling the upper one.

PARTICLES

Figure 2.3
Diffusion of bromine in air.

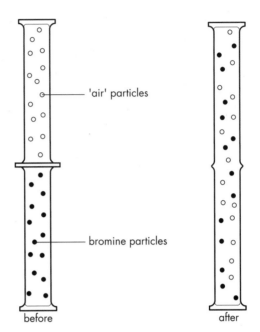

'air' particles

bromine particles

before   after

Materials
- eye protection
- 2 gas jars with covers
- Vaseline
- dropping pipette
- small bottle of bromine (or small glass ampoules)
- PVC gloves

Safety
- *Bromine has an irritant vapour. It causes severe burns and it is very toxic by inhalation.*
- *Always use bromine in a fume cupboard, wearing a lab coat, eye protection and PVC gloves. A beaker of dilute sodium thiosulphate solution should be available.*

Procedure
1. Smear the gas jar covers with Vaseline.
2. In a fume cupboard, take out a very small quantity of bromine from the bottle using the dropping pipette and put four or five drops into one gas jar.
3. Close the jar with a cover.
4. Invert a second 'empty' gas jar over the first and slide out the cover.
5. Watch the brown bromine colour diffuse from the lower into the upper gas jar. Ask pupils to explain the observations and get them to draw suitable 'pictures' to illustrate the movement of the particles.

Demonstration of diffusion of ammonia and hydrogen chloride

This experiment can be used to compare the rates of diffusion of ammonia and hydrogen chloride particles. Pads of cotton wool soaked in ammonia solution and in hydrogen chloride solution (called hydrochloric acid) are simultaneously put at opposite ends of a long horizontal glass tube (see Figure 2.4). Where the hydrogen chloride and ammonia gases meet, a white solid called ammonium chloride is formed:

ammonia	+	hydrogen chloride	→	ammonium chloride
$NH_3(g)$	+	$HCl(g)$	→	$NH_4Cl(s)$

Figure 2.4
Diffusion of ammonia and hydrogen chloride in air.

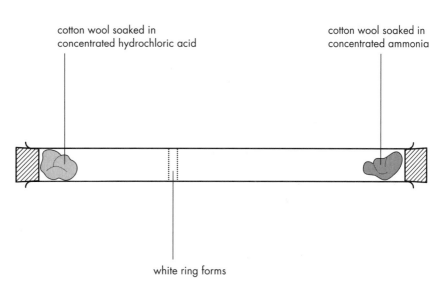

cotton wool soaked in concentrated hydrochloric acid

cotton wool soaked in concentrated ammonia

white ring forms

Materials
- eye protection
- length of glass tubing, about 2 cm in diameter and 1 m long (It is essential that this is completely dry. Heat it in an oven at 105 °C immediately before use.)
- 2 retort stands, with bosses and clamps
- 2 pairs of tongs or tweezers
- 2 corks to fit the ends of the glass tube
- black card (or other material) to act as a background
- PVC gloves
- cotton wool
- concentrated ammonia solution (18 mol/dm³)
- concentrated hydrochloric acid (10 mol/dm³)

<u>Safety</u>
- *Concentrated ammonia is corrosive and causes burns; its vapour is irritating to the eyes, the respiratory system.*
- *Concentrated hydrochloric acid produces an irritating vapour and the liquid is corrosive and can cause burns.*
- *A fume cupboard should be used.*
- *Wear eye protection when using ammonia and hydrochloric acid.*

<u>Procedure</u>
1. Demonstrate what happens when ammonia and hydrogen chloride gases meet: in a fume cupboard remove the stoppers from bottles of concentrated ammonia and hydrochloric acid (tell pupils that hydrogen chloride is the gas that comes out of the bottle) and gently blow across the open bottles. As the gases meet, white clouds form, containing solid particles of ammonium chloride.
2. Support the glass tube horizontally between two retort stands.
3. In a fume cupboard dip one piece of cotton wool in concentrated hydrochloric acid and another in concentrated ammonia solution.
4. Allow the surplus liquid to drain off.
5. Place the pieces of cotton wool in the glass tube and at the same time insert corks into the ends of the tube.
6. View the tube against a black background until the white solid is formed. (This takes about 5 minutes.)
7. Measure the distance of the ring from each end of the tube.

<u>What you might expect</u>
A disc of white solid appears when the gases meet (this will take some time). The white solid appears closer to the hydrogen chloride end. Encourage the pupils to suggest possible reasons for these two observations. If they have difficulties, ask where the ring would form if the ammonia and hydrogen chloride particles moved at the same speed. Eventually they should appreciate that the ammonia particles must move about twice as fast as the hydrogen chloride particles, as the ammonia particles travel about twice the distance.

Air particles are present in the 'empty' tube and collide with the other particles, slowing down their forward progress. A useful analogy here is to compare the situation in the tube with the movement of pupils along a crowded corridor. The ammonia 'particles' are lighter than the hydrogen chloride ones. Particle pictures should again be drawn.

Diffusion of carbon dioxide
In this experiment pupils are able to confirm the ideas obtained from the bromine diffusion demonstration.

Materials
- 4 test-tubes, 150 × 16 mm
- lime water
- source of carbon dioxide (cylinder or Kipp's apparatus)

Figure 2.5
Diffusion of carbon dioxide in air.

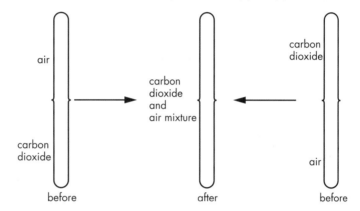

Procedure
1. Remind the class of the test for carbon dioxide gas by adding a few cm^3 of lime water to a test-tube filled with carbon dioxide and seeing it go cloudy. Also point out that carbon dioxide is more dense than air.
2. Fill two test-tubes with carbon dioxide with the tubes' mouths upwards (you can only really estimate when the tubes are full) and then cork them.
3. In the first experiment, place an 'empty' test-tube mouth to mouth on top of a carbon dioxide tube (see Figure 2.5).
4. In the second experiment, place the carbon dioxide test-tube on top of an 'empty' one. Encourage the pupils to speculate what will happen when the corks are removed. Pupils are often confused in this experiment and tend to refer to the higher density of the carbon dioxide rather than the nature of diffusion.
5. Remove the corks and hold the tubes in position for about 4 to 5 minutes.
6. Separate the test-tubes and test for the presence of carbon dioxide in each tube using lime water.
7. Finish the experiment by getting pupils to explain the results, using sketch diagrams of how the particles are moving and mixing.

What you might expect
Carbon dioxide should be found in each tube.

Diffusion in liquids
Materials
- eye protection
- 250 cm³ beaker
- tweezers
- potassium manganate(VII) crystals

Safety
- *Potassium manganate(VII) is an oxidising agent and is harmful.*
- *Potassium manganate(VII) crystals should never be touched with fingers as they cause brown stains.*
- *Wear eye protection when handling potassium manganate(VII).*

Procedure
1. Use tweezers to place a small purple crystal of potassium manganate(VII) in the bottom of a beaker of water.
2. Leave it for several days and watch the pink-purple colour spread.

What you might expect
Diffusion occurs slowly in liquids – much more slowly than in gases. Diffusion also takes place in solids but very, very slowly. With very able pupils you could explain how the rate of diffusion in solids could be found using radioactive isotopes.

Changes of physical state: melting solids and boiling liquids
If a solid is heated, the temperature–time graph, which is called the 'heating curve', can be drawn schematically as in Figure 2.6. Cooling curves are exactly the reverse of this.

Figure 2.6
A heating curve.

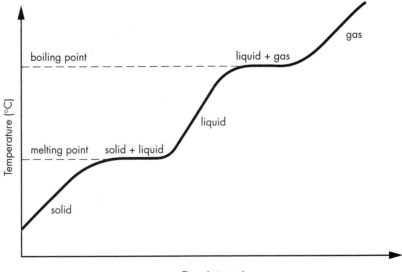

The effect of heat on a solid provides information about the bonding and structure of the substance but at this point the temperature–time graph is best explained in terms of what is happening to the arrangement and movements of particles and to the energy that is transferred.

The key points are that as a solid is heated the vibrations of the particles increase and the temperature increases. At a particular temperature, the bonds between the particles begin to break and the ordered arrangement in the solid breaks down. The solid melts to a liquid, the temperature remaining constant during this process. When all the solid has melted, the temperature rises again as the movement of the particles increases. At the boiling point most of the remaining bonds between the particles are broken, allowing the particles to spread apart. This process requires energy and so the temperature remains constant during this time.

Pupils continue to find it difficult to picture what is happening, and suitable particle sketches and models are vital. The fact that the bonds *between* particles are broken during melting and boiling and so the temperature remains constant generally causes the main problem for pupils.

The exact values of melting points and boiling points of substances depends on the particles that are present and on the forces between them. No two substances have the same melting and boiling points, and an impure substance has a lower melting point and a higher boiling point than the pure substance. Therefore, melting points and boiling points can be used to identify substances and can be used as an indicator of purity.

Recent National Tests showed that pupils had difficulty in determining the physical state of substances when given boiling point and melting point data. A simple exercise such as provided in the Nuffield Chemistry Activity sheet C2 *Solid, liquid or gas?* may be helpful here. The following activity provides 'hands on' experience of the phenomena.

Obtaining a heating curve

Materials

- eye protection
- boiling tube
- thermometer, -10 to $110\,°C$
- stopwatch
- beaker
- retort stand, boss and clamp

- tripod, gauze and heatproof mat
- Bunsen burner
- stearic acid (octadecanoic acid, m.p. 67–69 °C) or 1-hexadecanol (m.p. 54–56 °C)

Safety

- *Accidents sometimes occur when the thermometer is trapped in solid stearic acid. Do not try to remove the thermometer if it becomes trapped in the solid: heat the boiling tube containing the stearic acid until the acid has melted. Then the thermometer can easily be removed.*
- *Wear eye protection when using a Bunsen burner.*

Procedure

Figure 2.7
Determining the heating curve of stearic acid or 1-hexadecanol.

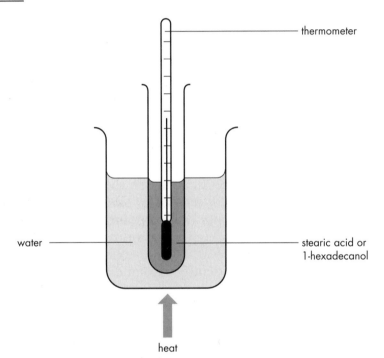

1. Put about 5 cm depth of the solid into the boiling tube (see Figure 2.7).
2. Put the thermometer in the solid and record the temperature.
3. Heat a beaker of water until its temperature is about 75 °C and try to keep this as constant as possible throughout the experiment.
4. Put the tube, with the thermometer still in the solid, into the beaker of hot water and start the stopwatch. The temperature will start to rise and it is a good idea for pupils to make their own judgements about when to record its value, e.g. if the temperature is rising rapidly then it should be recorded more often but if it is going up slowly then fewer measurements are needed. Note when changes in the physical state of the solid occur.

The pupils should record all the results in the most appropriate way and plot a graph to show how the temperature of the substance changed with time.

Pupils could be asked to draw any conclusions from the experiment but they will need prompts in the form of questions such as:

- What happens to the temperature of the stearic acid while it is melting?
- Energy is transferred from the hot water to the stearic acid during the experiment. What happens to the stearic acid particles (molecules) as
  a) the solid acid is warmed?
  b) the acid melts?
  c) the liquid acid is warmed?

This experiment provides a good opportunity for pupils to be taught or to improve their practical skills. Making accurate measurements and recording them clearly and accurately is a key feature of the 'obtaining evidence' skill. Plotting graphs, drawing lines of best fit and making appropriate conclusions are necessary parts of the 'analysing evidence' skill.

What you might expect

In an ideal case, the graph obtained should have the shape shown in Figure 2.6, page 56. In practice it is often difficult for pupils to obtain clear evidence of the plateau in the temperature–time curve which corresponds to the melting point.

Extension

This might be a good opportunity for some data-logging which you might like to run alongside the pupil experiment, with a thermometer probe and associated computer software. It is possible to record both the heating-up and the cooling-down process but it is probably a good idea to have the pupils plot graphs and answer questions because this data-logging exercise can take some time to get good results.

◆ *Enhancement ideas*

- ◆ A microscale study of gaseous diffusion (No. 32 in *Microscale Chemistry*, Royal Society of Chemistry, 1998, ISBN 1 870343 49 2).

2.2 Types of particle

Pupils generally understand that matter is made up of particles. Now they have to learn what these particles actually are and to use terms such as 'atoms', 'molecules' and 'ions' correctly.

♦ *Previous knowledge and experience*

Pupils are likely to have heard of atoms and molecules but they are unlikely to have a clear picture of what they are.

♦ *A teaching sequence*

Once the ideas of the particle theory have been firmly established, the three different types of particle that are the building blocks for all materials could be considered. One approach is to introduce the different types as and when they are needed, while another way would be to spend an initial period describing and discussing them all together. The latter approach provides opportunities to re-visit and reinforce when they are met again. This does have some appeal, particularly perhaps for less able pupils.

The essential points to make are that:

- *atoms* are single, neutral particles
- *molecules* are also neutral but are made up of two or more atoms chemically bonded together
- *ions* are positively or negatively charged particles.

You may find it helpful to refer back to the terms 'elements', 'mixtures' and 'compounds' in Chapter 1 in order to revise and clarify any difficulties of understanding. Pupils often confuse these terms.

At this stage an exhibition of models can be useful to emphasise the differences between atoms, molecules and ions. This exhibition could include models of neon (single balls), an oxygen molecule (two balls stuck together), carbon dioxide, methane and sodium chloride. In each model pupils should decide what the balls represent. In oxygen, for example, each ball represents an oxygen atom but the pair together represent an oxygen molecule. In sodium chloride, the balls represent ions: a single sodium chloride molecule does not exist. Hopefully, pupils will be able to use the words 'atom', 'molecule' and 'ion' correctly.

2.3 Atomic structure

◆ *Previous knowledge and experience*

Some pupils might well have come across the idea that atoms are divisible, having heard of 'splitting the atom' say, and made up of smaller particles. They might possibly even know the names of the main sub-atomic particles.

Some pupils might be familiar with the term 'isotope' and will often associate it with radioactivity and danger.

◆ *A teaching sequence*

Structure of the atom

A good way to start might be by putting into perspective the size of the atom in relation to other things in our universe (see Table 2.3): can pupils comprehend the small size of the atom?

Table 2.3 *Atoms are tiny compared with most familiar things!*

	Measurements in m
Distance to the Sun	100 000 000 000
Diameter of the Earth	10 000 000
Height of Mount Everest	10 000
Human	1.5
Grain of sand	0.000 1
Virus	0.000 001
Sugar molecule	0.000 000 001
Atom	0.000 000 000 1

Pupils will probably have little awareness of charge, relative mass and above all the overall scale within the atom.

A simple table and sketch will suffice as an introduction, drawing attention to the very small size of the nucleus relative to the overall size of the atom. Appropriate analogies can be useful here; for example, if the atom was the size of Wembley Stadium, the nucleus would be the size of a pea on the centre circle.

It is a good idea to emphasise that the nucleus is positively charged and contains most of the mass of the atom. It contains protons and neutrons (see Table 2.4).

Table 2.4 *The main sub-atomic particles.*

Particle	Charge	Relative mass
Proton, p	+1	1
Neutron, n	0	1
Electron, e	−1	Negligible

The arrangements of the electrons are best not developed in detail at this stage, since pupils might be overloaded with new information: represent or describe electrons simply, either as a 'cloud' around the nucleus or as a series of dots. The overall structure of the atom and the types and charges of the particles involved are the essential points.

Isotopes

Before introducing the term 'isotope', it is important to establish that an atom of an element must contain equal numbers of protons and electrons to ensure that it is neutral, i.e. that it has no overall charge. The number of protons (or electrons) determines which element it is. For example, any atom containing 1 proton and 1 electron must be hydrogen.

The definitions of the *atomic* or *proton number* (the number of protons in the nucleus) and the *mass number* (the number of nucleons, i.e. protons and neutrons in the nucleus) could be introduced now. Using either a Periodic table or a list of elements, pupils could find atomic and mass numbers for various atoms. They could work out the numbers of protons, electrons and neutrons in each. For example lithium, with atomic number 3 and mass number 7, contains 3 protons, 3 electrons and 4 neutrons.

Pupils should notice that:

- the larger the atom, the greater the proportion of neutrons; for example, helium contains 2 protons and 2 neutrons and uranium contains 92 protons and over 140 neutrons
- some elements have more than one type of atom, containing different numbers of neutrons; for example, hydrogen has three isotopes, all containing 1 proton and 1 electron but with 0, 1 or 2 neutrons.

Pupils should now be introduced to the term *isotope*: isotopes are atoms of the same element containing different numbers of neutrons. (Pupils frequently confuse the word 'isotope' with 'allotrope' and with 'isomer'.)

They could now use the following notation to represent isotopes such as hydrogen-1, hydrogen-2 and hydrogen-3:

$$\text{mass number } \atop \text{atomic number}} X$$

The three isotopes of hydrogen can then be represented as:

$$^{1}_{1}H \quad ^{2}_{1}H \quad ^{3}_{1}H$$

It is important that pupils remember that the smaller number is always at the bottom. Now they could represent the three isotopes of oxygen – oxygen-16, oxygen-17 and oxygen-18 – in a similar way.

At this stage, the term *relative atomic mass* could be discussed. The relative atomic mass was originally taken as the ratio of the mass of an atom to the mass of the lightest atom, a hydrogen atom. As a hydrogen atom was the lightest atom, all relative atomic masses would be greater than 1. Only with more able pupils might you explain why carbon-12 is now taken as the standard.

Looking at a list of relative atomic masses, for example Table 2.5, will reveal some interesting points:

- the relative atomic mass of most elements is close to, but not exactly, a whole number
- some, such as chlorine, have a relative atomic mass that falls clearly away from any whole number.

Table 2.5 *Some relative atomic masses.*

Element	Relative atomic mass
Hydrogen	1.0080
Oxygen	15.999
Nitrogen	14.0067
Sodium	22.989
Chlorine	35.453

Calculations

The term 'relative atomic mass' can be discussed in a little more detail with able pupils and the idea of weighted averages of the masses of the isotopes introduced.

For example, chlorine is made up of two isotopes, one of mass number 35 and the other of 37. Approximately 75% is $^{35}_{17}Cl$ and approximately 25% is $^{37}_{17}Cl$. Therefore, the relative atomic mass of chlorine is

$$\frac{75}{100} \times 35 + \frac{25}{100} \times 37 = 35.5$$

Neon, which is made up of three isotopes, provides a further example. 90.9% is neon-20, 0.26% is neon-21 and 8.8% is neon-22. The relative atomic mass is 20.179.

The idea that, owing to the existence of isotopes, in most cases no atom of a particular element has the same mass as the value of its relative atomic mass takes most pupils by surprise.

Table of elements

You might like to start to draw up a table showing the detailed atomic structure of each element in turn, as in Table 2.6. Once the pattern is established and understood, you might get each pupil to look up appropriate data using either a data book or a software package and then contribute to the completion of the table for the first 20 elements.

Table 2.6 *The atomic structures of the first five elements.*

Element	Symbol	Atomic number	Number of protons	Number of electrons	Mass number of main isotope	Number of neutrons in main isotope
Hydrogen	H	1	1	1	1	0
Helium	He	2	2	2	4	2
Lithium	Li	3	3	3	7	4
Beryllium	Be	4	4	4	9	5
Boron	B	5	5	5	11	6

Electronic structure

The emphasis now moves to looking at the arrangement of electrons in an atom, or its 'electronic structure'. It is important to remind pupils regularly that atoms are not charged overall because the amount of positive charge due to protons balances out the negative charge of the electrons.

While the nucleus is of interest to the nuclear physicist, Chemistry is all about the arrangements and interactions of electrons and how they determine physical properties and chemical changes.

Electrons can be considered to be arranged in various shells or energy levels well outside the nucleus. A simple sketch is appropriate here and it is probably not a good idea to go into much more detail unless pushed: while providing an instant answer, describing this picture in terms of electrons rotating around the nucleus rather like the planetary system can cause difficulties later on.

The electronic structure of each element can be developed by reference to the Periodic table, pointing out that each adjacent element has one more proton and therefore one more electron in its structure. When a row or period is completed, a new shell is started. There is a maximum of two electrons in the first shell, eight in the second and eight in the third.

Electron shells

Using a Periodic table format, sketches of the electronic structures of the first few elements can be drawn by pupils and the teacher in collaboration and filled with the appropriate numbers of electrons. To avoid boredom due to repetition, it is probably best with further elements (up to about 20), for pupils either to complete pre-drawn electronic shells on a worksheet (Nuffield Chemistry Activity sheet C82) or to write out the electronic structures in abbreviated form, e.g. 2, 8, 1.

♦ *Further activities*

♦ Electronic structure can also be introduced by using experimental evidence, and an alternative approach involving ionisation energies taken from a data book or software package is also a possibility for more able pupils. You might like to organise this activity as a teacher–class collaboration.

♦ The Periodic table is closely linked with electronic structure. Chapter 7 describes these links and relates trends in reactivity with electronic structure.

♦ Reference to a list of isotopes of all the elements shows that some elements have many stable isotopes but some have only one. Which elements have only a single isotope? Is there any pattern in this?

2.4 Bonding

♦ *Previous knowledge and experience*

Pupils are unlikely to have come across ideas of bonding between atoms.

♦ *A teaching sequence*

The focus now moves to the study of what holds the particles together in elements and compounds, i.e. bonding. A suitable starting position is to remind pupils that most elements can be simply classified into two types, metals and non-metals. Therefore it might be expected that only three different types of bonding exist, for the combinations:

- metal with metal – metallic bonding
- metal with non-metal – ionic bonding
- non-metal with non-metal – covalent bonding.

Of course, many compounds have more than one type of bonding within them, for example copper sulphate has covalent bonding within the sulphate ion and ionic bonding between the copper and the sulphate ions. Furthermore, none of the types of bonding described above are present in the noble gases, and care must always be taken with such simplified but useful descriptions.

Metallic bonding

A fairly simple picture is all that is needed at this level. The idea that the outermost electrons of each atom can spread or 'delocalise' across all the other atoms in the structure, such that they become attracted to the positively charged nuclei and therefore hold the structure together, is all that is needed. This 'sea' or 'cloud' of electrons provides the conceptual model needed to explain the key properties of electrical and thermal conductivity of metals and alloys.

Pupils can make a model of a layer of atoms in a metal by making a triangle with three books and fitting as many marbles (or polystyrene balls) into the triangle as possible. The close packing that results produces an arrangement in which each atom in the layer is surrounded by six others in the form of a regular hexagon. Now ask pupils how many atoms would be around any atom inside the structure when a series of layers is placed one on top of another. The answer is 12 – three in the layer above, six in the same layer and three in the layer below. Pupils may realise that the close packing of atoms in a metal structure leads to the typical high density of metals.

Ionic bonding

The second type of bonding occurs in compounds of metals and non-metals and is called ionic bonding. It is important to stress this, and a good idea is to ask pupils to give examples of metal–non-metal compounds in a 'round the class' question-and-answer session.

The essential idea for pupils to appreciate is that in ionic bonding metals transfer electrons to the non-metals such that the nearest noble-gas electronic structure is obtained for both the metal and the non-metal. (Transition metal ions, e.g. Fe^{2+}, are an exception.) The consequence of this electron transfer is that the metal atoms become positively charged ions and the non-metals become negatively charged ions. It is these outcomes which pupils find very difficult to understand, particularly the formation of positive ions.

Appropriate pictorial representations are crucial in this regard. One way which has been found helpful is to represent the process just in terms of the *numbers* of particles involved, i.e. protons and electrons, as in Figure 2.8 for the formation of sodium chloride.

Figure 2.8
The formation of sodium chloride, shown in terms of the particles involved.

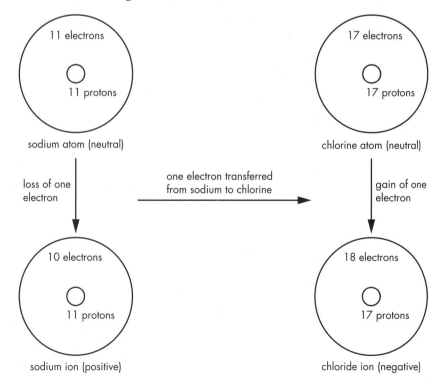

11 electrons

11 protons

sodium atom (neutral)

17 electrons

17 protons

chlorine atom (neutral)

loss of one electron

one electron transferred from sodium to chlorine

gain of one electron

10 electrons

11 protons

sodium ion (positive)

18 electrons

17 protons

chloride ion (negative)

This can then be summarised as:

sodium atom   →   sodium ion (positive) + 1 electron

Na   → Na^+ + e^-

chlorine atom +   1 electron   → chloride ion (negative)

Cl   + e^- → Cl^-

This can be followed up by the more usual pictorial representations showing the electron shells with electrons as dots and crosses (see Figure 2.9).

Figure 2.9
The formation of sodium chloride, shown as a 'dot-and-cross' diagram.

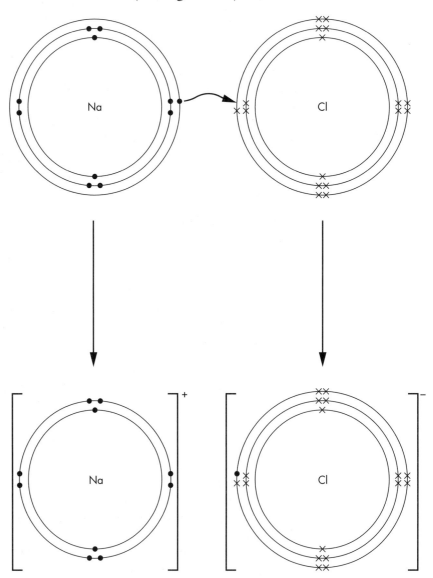

A common misconception by pupils is that the electrons in different elements are actually different, and it is important to point out that the dot-and-cross representation is only used to help keep a track of the number of electrons and where they have come from.

It is important that pupils are given plenty of practice in this area, including the situation where two electrons are transferred from the metal atom, e.g. in magnesium fluoride (MgF_2) and calcium oxide (CaO).

It is important to ensure that the pictorial representations of ions clearly indicate the separate nature of the species involved. Sometimes putting 'square brackets' with the appropriate charges around each ion emphasises this point and avoids any confusion (in, say, an examiner's mind) with covalent bonding.

It also helps for future work to point out that the attraction between the positive and negative ions extends throughout the whole structure and that it is this that really represents 'ionic bonding' and holds all the ions together.

Pupils generally find this section quite difficult, particularly identifying the types of particle and bonding present and drawing clear dot-and-cross diagrams.

Covalent bonding

The bonding between non-metals atoms is invariably covalent and involves the sharing of electrons such that each atom involved in the covalent bond achieves the nearest noble-gas electron configuration. The so-called 'octet rule', although useful in the introduction to this area, does have serious limitations and perhaps should not be over-emphasised.

There are a number of ways of representing the formation of covalent bonds:

- pictorial representations showing all the shells of electrons
- pictorial representations showing only the outermost electronic shell
- representations showing only the outermost electrons
- 'bond type' diagrams.

It can be rather confusing for pupils if different ways of showing covalent bond formation are used, and it is probably better to adopt one way which suits the level of ability of the pupils being taught.

Hydrogen, chlorine, hydrogen chloride, water and carbon dioxide molecules provide an appropriate start (see Figure 2.10).

Figure 2.10
Representations of covalent bonding.

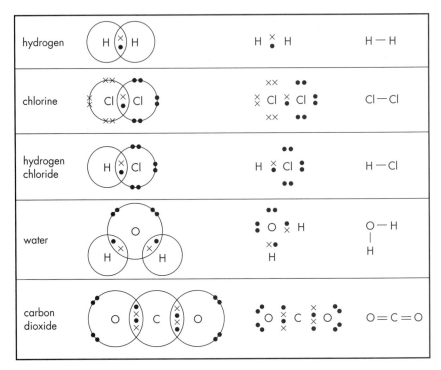

Again, plenty of practice is needed if pupils are to fully understand the nature of such diagrams. The fact that covalent bonding is directional between two atoms is also a point worth noting for able students.

♦ *Enhancement ideas*

♦ Just as black paint and white paint are extremes, with all shades of grey as intermediates, ionic and covalent bonding are extremes and all sorts of intermediate bonding are possible. The Pauling electronegativities of potassium and fluorine are 0.8 and 4.0; the difference in electronegativity is 3.2, corresponding to 92% ionic character. Lithium iodide is about 43% ionic in character. Use a data book such as *Chemistry Data Book*, 2nd edition in SI (1999) by Stark, J.G. and Wallace, H.G., London, John Murray, to estimate ionic character in other compounds.

♦ Pupils could investigate ways in which layers of metal atoms can be packed – ABABAB and ABCABC. Alkali metals have lower densities than most metals: they float on water. Pupils could research structures of metals. They will find that atoms in alkali metals do not have typical metallic structures.

2.5 Structure

♦ *Previous knowledge and experience*

Pupils are unlikely to have come across ideas of the structures of chemical substances, although they will have probably seen pictures of molecular structures, such as that of DNA.

♦ *A teaching sequence*

Particles are held together in substances by what is referred to as 'bonding', but their arrangement is described in terms of their 'structure'. Structure and bonding are key areas in Chemistry and provide a most important explanatory framework for the properties and reactions of substances. It is the structure of a substance which will determine whether it is soft, hard, fibrous, of high or low melting point or elastomeric. The properties of a substance give information about its probable structure. It is important to stress the interconnection between structure and properties at regular intervals.

Types of structure – giant and molecular

It is possible to divide substances in the solid, liquid or gaseous states into two main types, giant and molecular. Some texts refer to 'giant molecules' such as diamond, on the grounds that the bonding is covalent in both small molecules and diamond. Others call polymers 'giant molecules'. At this stage it seems better to avoid the term and separate clearly the words 'giant' and 'molecule'.

In a giant structure there is a continuous two- or three-dimensional network of particles, atoms or ions, connected by strong bonds throughout. Substances with giant structures are found only in the solid state and they invariably have high melting points, which distinguish them clearly from substances with molecular structures. Examples include diamond, graphite, metals and metal salts.

Substances with molecular structures contain distinct separate units, or molecules. These can be simply divided into those with small molecules, containing a relatively few atoms bonded together, and those with large molecules involving thousands of atoms. The bonding holding the atoms together within the molecule is strong, but the bonding between the molecules is relatively weak in most cases. It is the combination of these two points that pupils continue to find difficult.

Boiling water provides an illustration that may be helpful here: the steam that is formed when water boils still contains water molecules, just further apart. A breakdown into hydrogen and oxygen does not occur.

Substances which are made up of small molecules are gases, low boiling-point liquids or low melting-point solids. Examples include oxygen, hydrogen, carbon dioxide, iodine and small organic molecules. Substances with large molecules are solids and can also be relatively easy to soften or melt, although often they decompose on heating, e.g. plastics.

Summary tables at appropriate times help pupils to construct suitable frameworks on which to support their developing knowledge.

Just a word of warning in this area: not all substances fit neatly into a particular slot in a particular pattern, by any means.

Use of models

The use of models of various types is crucial if pupils are to clearly appreciate the different types of structure and to produce their own mental pictures of them. The ball-and-stick types are probably the most convenient and useful to illustrate the arrangement of particles. However, you must guard against giving pupils the false impression that the space between atoms is unoccupied, and occasional space-filling models are good to help avoid this confusion. Furthermore, there are a number of suitable computer programs which can provide illustrative support in this area, either for demonstration purposes or individual pupil 'hands on' experience, depending on the facilities available. Overhead projector transparencies are also available.

Permanent structural models of the common substances with giant structures, such as diamond, graphite, sodium chloride and metals, all help pupils to see arrangements of particles in three dimensions clearly.

It is important to allow pupils some 'hands on' experience of building models, and it is probably best to limit this to simple molecular structures using the ball-and-stick models. Distribution of such model-building kits to a class of pupils is not without its problems, and one way is to divide the appropriate number of balls and sticks into individual packs in small plastic bags.

At the end of the day it is still important that pupils have experience of drawing freehand pictures of some simple structures, e.g. water and methane, and perhaps a small part of a giant structure, such as graphite, diamond or sodium chloride (see Figure 2.11).

Figure 2.11
Ball-and-stick models of some giant structures.
a Graphite.
b Diamond.
c Sodium chloride.

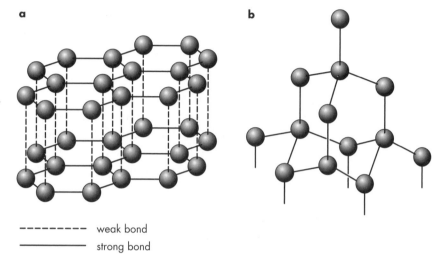

a   b

--------- weak bond
————————— strong bond

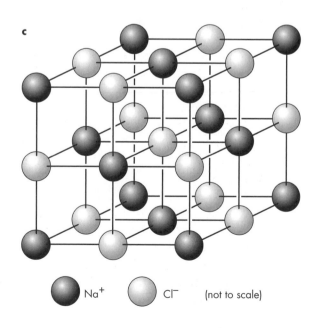

c

Na$^+$ Cl$^-$ (not to scale)

It might be worth mentioning at the start that the structures of solids can be determined by the technique of X-ray diffraction, developed by the father-and-son team of Sir William and Sir Lawrence Bragg in the early 20th century. They found that there was a simple law connecting the wavelength of X-rays and the spacing of atoms in crystals. X-rays are waves of the same nature as visible light but with a wavelength that is 10 000 times shorter. This is of the same order as the distances between particles in the solid state.

Effect of heat on substances

One way to continue the topic is to investigate the effects of heat on substances and develop the ideas of strength of bonding between particles. This can help pupils to understand that substances can be classified by structure and the use of the terms 'giant' and 'molecular'.

Heating some solids

Materials
- eye protection
- tin lids
- tongs
- Bunsen burner
- tripod, gauze and heatproof mat
- Pyrex test-tube
- PVC gloves

Chemicals for pupils – have small samples of the following solids arranged in suitable groups:

- A: copper, iron, zinc
- B: graphite, silicon, sand
- C: sodium chloride, magnesium oxide, calcium carbonate
- D: candle wax, glucose, polythene

Chemicals for teachers – the following samples should be heated gently in a fume cupboard because they are flammable and/or toxic:

- sulphur
- iodine
- polystyrene

<u>Safety</u>
- *Sulphur is a fine powder which irritates the eyes.*
- *Sulphur vapour is flammable, forming sulphur dioxide gas. Sulphur dioxide is toxic and corrosive; it may also trigger an asthma attack in a sensitive individual.*
- *Use a fume cupboard when heating sulphur or iodine. However, if a mineral wool plug is used in the test-tube, then sulphur or iodine can be heated in the open laboratory.*
- *Iodine is harmful by inhalation and skin contact and causes burns over time.*
- *Iodine vapour crystallises very painfully on the eyeballs.*
- *Polythene and polystyrene can give off toxic gases if heated too strongly.*
- *Use a fume cupboard when heating polythene and polystyrene.*
- *Wear eye protection when heating polythene and polystyrene.*
- *Wear eye protection when using a Bunsen burner.*

Pupils should heat at least one solid from each group, perhaps two from group D.

<u>Procedure</u>
1. Place a small sample of the solid on a tin lid.
2. Heat the sample gently. If anything happens, stop heating and note down the observations.
3. If nothing appears to happen, continue heating more strongly and record any observations.
4. Repeat the experiment with other solid samples.

<u>Demonstration</u>
The teacher then demonstrates heating the remaining samples in the fume cupboard. Ask the pupils to tabulate their results, writing the name of each substance tested and the observations made on heating.

Ask them to suggest what sort of strength (strong or weak) the forces must be between the particles in each substance.

Introduce the terms 'giant structure' and 'molecular structure' and identify the type of structure for each solid.

Classify each solid in terms of whether it is a metal element, a non-metal element, a compound of a metal and a non-metal or a compound of two non-metals. Ask pupils to suggest simple guidelines for predicting whether a substance will have a molecular or a giant structure. Do they know of any exceptions to the general guidelines?

<u>What you might expect</u>
In fact, out of the chemicals provided for the pupils nothing much will happen with the solids in groups A, B and C, although beware of water vapour coming off which might be confusing. These substances all have giant structures.

Substances in group D and also those in the demonstration experiment will be affected by heat fairly easily. These substances have molecular structures.

Candle wax, polythene and polystyrene will melt or soften to colourless liquids if heated gently enough, as will glucose, although that is more likely to decompose to a brown–black sticky mess.

The sulphur will first melt to a mobile yellow liquid but if the heating is continued the liquid will darken and become more viscous.

The iodine will melt or sublime very quickly, and toxic purple fumes will form immediately.

Electrolysis and solubility

You might like to bring together other aspects of the course, such as electrolysis and solubility, either to finish the topic or for revision purposes. A useful reference is Nuffield Science Activity sheet C86 (1996).

Metals and graphite conduct electricity without decomposition because they allow the flow of free electrons through their structure. Substances with giant ionic structures of course cannot conduct in the solid state because the ions cannot move. However, when molten or when dissolved in water the ions are free to move and electrolysis occurs with decomposition of the substance. It is the difference between these two types of process which often confuses all but the ablest pupils.

Substances with molecular structures do not conduct in any state (except in those cases when a chemical reaction has occurred, e.g. when hydrogen chloride dissolves in water to form ions).

Solubility is a complex area and probably should only be covered for able pupils. However, the ideas which can be covered at this level are that the solubility of a solute in a solvent will depend on the strength of the forces between the particles in the solute and the solvent themselves, on the one hand, and the attractive forces between the solute and solvent particles together on the other. The fact that many ionic substances dissolve in water suggests that there is a strong attraction between the ions and water molecules.

Investigating the properties of hexadecanol and potassium iodide

This activity allows the opportunity for the ideas and concepts met earlier to be consolidated and revised. It is a combination of teacher demonstration and pupil activity.

Materials
- eye protection
- test-tubes, 125 × 16 mm
- 6–9 V battery or transformer
- leads with crocodile clips
- graphite electrodes
- lamp
- hexadecanol
- potassium iodide
- cyclohexane

 Safety
- *Cyclohexane is highly flammable, do not have flames near when using it.*
- *Wear eye protection when heating the two substances.*

Give pupils an outline table, similar to Table 2.7, to record their observations and answers to the discussion questions.

Table 2.7 *Investigating the properties of hexadecanol and potassium iodide. (Adapted from* Chemistry *by J.A.Hunt & A.Sykes, 1986. Longman, Harlow.)*

Property	Procedure	Discussion
Appearance	What is their physical state?	What does this suggest about the arrangement of particles in this substance?
Smell	Can you smell the material?	What does this tell you about the strength of bonds between the particles in the substance?
Effect of gentle heating	Place each substance in a test-tube and heat in a beaker of boiling water	What does this tell you about their melting points and the energy needed to separate their particles?
Does the pure substance conduct electricity?	(Teacher demonstration) Set up an electrolysis circuit using a battery, graphite electrodes and a lamp	Are the pure substances conductors? What does this tell you about the nature of the particles in them?
Solubility in **a)** water and **b)** cyclohexane (possible teacher demonstration)	**a)** Add one crystal of the pure substance to 5 cm^3 of water in a test-tube. Warm gently in hot water (from an electric kettle) and shake if needed. Repeat with the other substance. **b)** (Teacher demonstration) Do the whole experiment again using cyclohexane instead of water, but this time *do not* have any flames near	Which is the better solvent for each solid?
Does the solution of the substance in water conduct electricity?	Pupils set up a simple electrolysis circuit as above	What does this tell you about the solutions?

Particles, bonding and structure: a summary

Pupils could fill in a blank copy of the flow chart in Figure 2.12 as a useful summary of the topic.

Figure 2.12
Particles, bonding and structure – a summary.

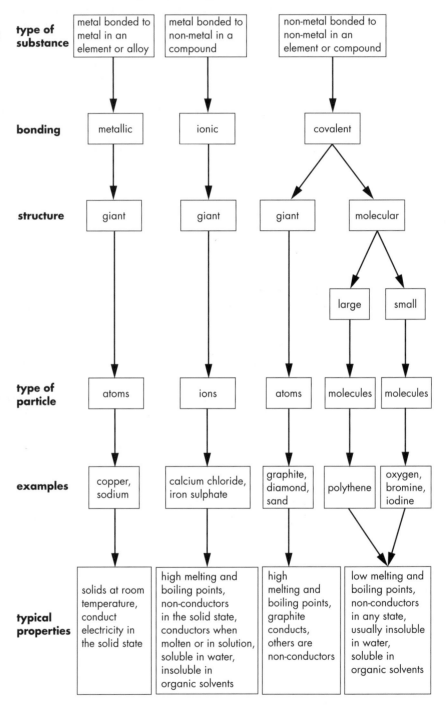

◆ *Other resources*

 Pupils are used to seeing diagrams of particles in solids, liquids and gases (see Figure 2.2, page 50). However, they sometimes fail to appreciate the differences in the movement of the particles in the three states. Simple animations of the movement of these particles can be found in the *Letts Revision Chemistry* CD ROM (available from Letts Educational, Aldine House, Aldine Place, London WC12 8AW). Included among other animations are particle movements during diffusion.

 Atoms, isotopes and atomic structure are clearly explained in the CD ROM *The Elements* from Yorkshire International Thomson Multimedia Ltd, Kirkstall Road, Leeds LS3 1JS.

Further possible practical work for this topic may be found in *Nuffield Science Activities for GCSE: Chemistry*, Nuffield Foundation (1996).

 Web sites

◆ Pupils can 'look inside the atom' on a series of web pages from the American Institute of Physics on:
www.aip.org/history/electron/jjhome.htm

◆ An elementary treatment of structure and bonding can be found on:
www.chem4kids.com/chem4kids/atoms/index.html
www.wilson.uscd.edu/education/xraydiff/xraydiff.html

◆ Pupils frequently believe that the idea of an atom as an indivisible unit of matter was forgotten after Democitus until Dalton in the 19th century. This is not the case:
http://dbhs.wvused.k12.ca.us/
Democritus-to-Dalton.html
explains how the ideas developed in between.

3 Chemistry of carbon compounds

David Lees

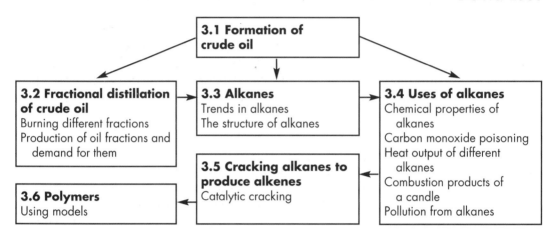

3.1 Formation of crude oil

3.2 Fractional distillation of crude oil
Burning different fractions
Production of oil fractions and demand for them

3.3 Alkanes
Trends in alkanes
The structure of alkanes

3.4 Uses of alkanes
Chemical properties of alkanes
Carbon monoxide poisoning
Heat output of different alkanes
Combustion products of a candle
Pollution from alkanes

3.5 Cracking alkanes to produce alkenes
Catalytic cracking

3.6 Polymers
Using models

◆ Choosing a route

Our use of chemicals obtained from crude oil as fuels for transport and heating requirements, and to provide the materials for the production of a wide range of substances, has increased dramatically over the last century. The study of carbon compounds, or 'organic chemistry', has become an important aspect of science courses.

Study may begin with some details of how crude oil (petroleum) was formed, or this may be left until later in the topic when pupils have gained some idea of what crude oil is.

A look at the fractional distillation of crude oil leads naturally to a discussion of the uses of the fractions obtained. Laboratory work can be linked to industrial methods and the needs of modern society.

The structures of the alkanes can then be related to their boiling points and their uses.

A study of the combustion of alkanes allows the topic to explore the energy changes associated with chemical reactions and the environmental impact of our use of alkanes as fuels.

The use of catalytic cracking to provide more of the most-used fractions from crude oil gives further opportunity for linking laboratory practicals with industrial processes.

The formation of alkanes in the cracking process can be developed into a consideration of polymer chemistry, with the opportunity to link properties with uses.

3.1 Formation of crude oil

♦ *Previous knowledge and experience*

Pupils may have some ideas from primary science of how coal was produced over millions of years by the action of high temperatures and pressures on plant material. The processes that produced crude oil (petroleum) were very similar, but the starting material was small animals.

♦ *A teaching sequence*

The topic of the formation of crude oil, exploration for oil and oil drilling and transportation may be covered by the Geography department, and it is important to liaise with the Geography department to ensure a coherent approach.

Pupils need to be able to describe the steps in the formation of crude oil. With lower ability pupils, write down the different steps on pieces of acetate before the lesson. Discuss the correct order of the steps with pupils.

Crude oil was formed from the remains of tiny animals which sank to the bottom of the sea when they died. The remains of these living organisms formed muddy sediments, where the water had little or no dissolved oxygen. Bacteria fed on the remains of the sea creatures, removing oxygen from the compounds from which the organisms were composed. Over a very long period of time, the remains were buried far below the Earth's surface. Here they were subjected to high temperatures and pressures, in the absence of air. This formed the black, viscous liquid we call 'crude oil'.

The liquid oil flowed up through porous rocks, such as sandstone, until further travel was prevented by impervious rocks. Oil deposits are therefore found under a cap of non-porous rock, such as shale (Figure 3.1, overleaf). The crude oil is trapped as tiny droplets in the pores between grains of sedimentary rock, which acts like a sponge. Natural gas is often found along with the crude oil.

When an oil well is sunk down to these deposits, oil and gas are forced to the surface under pressure. You can show pupils pictures of gas being burned off at oil wells. Oil is transported by pipeline or tanker to an oil refinery where the oil is turned into useful products.

Figure 3.1
*Deposits of gas
and oil trapped
under a cap of
non-porous rock.*

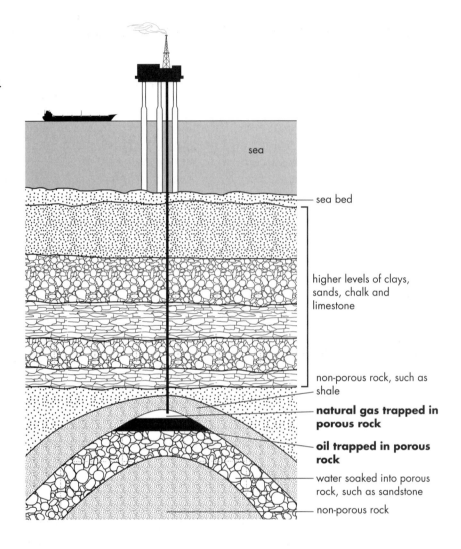

sea

sea bed

higher levels of clays,
sands, chalk and
limestone

non-porous rock, such as
shale

**natural gas trapped in
porous rock**

**oil trapped in porous
rock**

water soaked into porous
rock, such as sandstone

non-porous rock

Crude oil (petroleum) is a mixture of a large number of
hydrocarbons – compounds containing hydrogen and carbon
only. It is important that pupils realise the importance of the
word 'only' here. Without it, sugar (a compound of carbon,
hydrogen and oxygen) would be considered a hydrocarbon.

♦ *Enhancement ideas*

♦ The formation of crude oil and the type of situation in
which it is now found give an opportunity for more detailed
study of the geological factors involved. Consideration
could be given to the exploration for and discovery of
crude oil.

3.2 Fractional distillation of crude oil

◆ *Previous knowledge and experience*

The processes of distillation and fractional distillation are covered in Chapter 1. Briefly, distillation is used to separate a liquid from non-volatile materials, e.g. water from sea water, and fractional distillation is used to separate mixtures of liquids with different boiling points. A demonstration of the small-scale fractional distillation of 'crude oil' (page 17) may be used here.

◆ *A teaching sequence*

Burning different fractions

After a revision of the processes of distillation and fractional distillation, possibly with a demonstration, it is important to explain that fractional distillation is one of the processes that takes place in an oil refinery. Note that genuine crude oil cannot be used in schools, as it contains benzene. Chapter 1 (page 17) contains information on making a substitute.

Because of the complicated nature of crude oil, it is impossible to produce pure compounds from the process. Different fractions are produced, and each has a definite range of boiling points and different physical properties.

The fractions can be burned in turn to show that the fractions with low distillation temperatures are the most volatile.

Demonstration of the burning of fractions

Materials

- eye protection
- white tiles
- tripod, gauze and heatproof mat
- splints
- mineral wool
- fractions obtained from the demonstration on page 17, or small samples of hydrocarbons, e.g. hexane and decane

Safety
- All petroleum ethers are highly flammable.
- 60–80 °C petroleum ether is harmful.
- Hexane and decane are highly flammable.
- The experiment must be carried out in a well-ventilated room or fume cupboard.
- A demonstration on an open bench should be surrounded by safety screens.
- Use small samples, only about 10 drops.
- Keep other flammable liquids well away from this demonstration.
- Wear eye protection when burning the samples.

Procedure

Figure 3.2
Burning fractions of 'crude oil'.

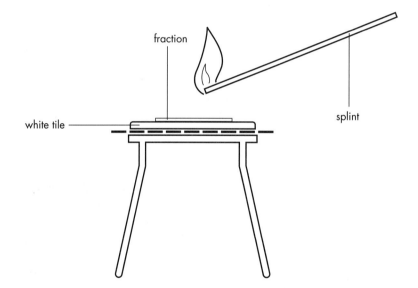

1. Pour a little of the fraction on to a white tile.
2. Slowly bring a lighted splint down towards the fraction on the tile (see Figure 3.2).
3. Repeat with the other fractions, using a fresh white tile for each.

As an alternative, the fractions can be poured on to a tuft of mineral wool.

What you might expect
Pupils will observe that the first fraction (petrol) burns when the splint is some few centimetres above the tile. As the other fractions are burned the splint needs to be increasingly close to the tile, until it proves impossible to get the fourth fraction to burn.

The laboratory distillation of simulated crude oil produces fractions which approximate to petrol, paraffin, diesel, lubricating oil and fuel oil. These act as an introduction for pupils to the fractions obtained in an oil refinery, which are listed in Table 3.1.

Table 3.1 *Fractions of crude oil from an oil refinery.*

Fraction	Boiling point range (°C)	Number of carbon atoms	Uses
Petroleum gases	Below 0	1–4	Bottled gas, chemical industry
Petrol (gasoline)	0–100	5–8	Fuel for cars
Naphtha	100–170	9–10	Chemical industry
Paraffin (kerosine)	170–250	11–14	Fuel for homes and aircraft
Diesel oil	250–340	15–19	Fuel for cars and heavy vehicles
Lubricants and waxes	340–500	20–35	Lubricants and polishes
Fuel oil	Over 500	36–45	Fuel for ships, electricity generation
Bitumen	(Not distilled)	Over 45	Road making, roofing

Production of oil fractions and demand for them

The quantities of the different fractions produced by primary fractional distillation depend upon the source of the crude oil. Table 3.2 gives the percentages of the different fractions obtained from crude oil from the North Sea and from the Middle East. It also gives the product demand by customers for the different fractions.

Table 3.2 *The mismatch between the production of fractions from crude oil and demand for them.*

Number of carbon atoms	Percentage in North Sea oil	Percentage in oil from the Middle East	Product demand as percentage
1–4	2	2	4
5–8	8	5	22
9–10	10	9	5
11–14	14	12	8
15–19	21	17	23
Over 20	45	55	38

Some time can be spent with pupils in looking at Table 3.2. You should be able to explain that the quantities of the different fractions obtained do not match the demand: there is a shortage of hydrocarbons with one to four carbon atoms (petroleum gases) and with five to eight carbon atoms (petrol). There is more of higher boiling point fractions. Also, it is possible to explain why North Sea crude oil is sold for a higher price than crude oil from the Middle East: it produces fractions in proportions closer to those demanded.

At this stage it is worthwhile mentioning that any process which could be devised to use up the fractions with lower demand could be economically important. This will lead later to the process of cracking.

◆ *Enhancement ideas*

- ◆ The distillation of crude oil could be extended to include consideration of the design of fractionating columns and how the fractionation process works.
- ◆ When fractions are collected by fractional distillation of 'crude oil' they can be tested for differences in viscosity and colour. The viscosity of liquids can be compared by timing how long a bead takes to fall a certain distance through the liquid.

◆ *Further activities*

- ◆ *Classic Chemistry Demonstrations* contains the following demonstration which is relevant to this section:
  - 61   Identifying the products of combustion
- ◆ *Microscale Chemistry* contains the following experiment which is relevant to this section:
  - 59   The treatment of oil spills

3.3 Alkanes

◆ *Previous knowledge and experience*

Pupils will not have studied the reactions of alkanes before.

◆ *A teaching sequence*

Trends in alkanes

Most of the compounds in crude oil are alkanes. These are hydrocarbons with the general formula C_nH_{2n+2}. They form a family of compounds with similar chemical properties, called a *homologous series*. In a homologous series, each member differs from the one before by the same grouping of atoms. In the case of the alkanes, each has one carbon atom and two hydrogen atoms more than the previous member of the series.

Table 3.3 shows the names, formulae (molecular and displayed) and boiling points of the first six alkanes.

Table 3.3 *The first six alkanes.*

Name of alkane	Molecular formula	Displayed (structural) formula	Boiling point (°C)
Methane	CH_4		−161
Ethane	C_2H_6		−88
Propane	C_3H_8		−42
Butane	C_4H_{10}		−1
Pentane	C_5H_{12}		36
Hexane	C_6H_{14}		69

If pupils are shown data for the first four alkanes, they can make predictions

- of the molecular formulae of the alkanes containing 20 carbon atoms ($C_{20}H_{42}$) and 35 carbon atoms ($C_{35}H_{72}$)
- about the molecular formula, displayed formula and boiling point of the next two members of the homologous series – pentane and hexane (see Table 3.3 – pupils will not be able to be exact with boiling points but should see a trend).

The structure of alkanes

The carbon and hydrogen atoms in alkane molecules are held together by single covalent bonds. These bonds, within the molecules, are strong; but the bonds between the alkane molecules are much weaker.

Considerable energy is needed to break an alkane up into hydrogen and carbon atoms, but relatively little energy is needed to separate the alkane molecules from each other. This explains why the smaller alkanes are gases at room temperature, and most of the larger ones are volatile liquids.

Pupils can get an idea of what alkane molecules are like by making models. This can be done using ball-and-spring or ball-and-stick models. Models of the first four members of the homologous series should be made to highlight an important difference between structural formulae and the true shapes of the molecules: the four bonds on each carbon atom actually point to the corners of a tetrahedron. This can be seen most clearly in a model of the methane molecule (see Figure 3.3).

Figure 3.3
Ball-and-spring model of a methane molecule.

The displayed formulae of ethane, propane and butane
(Table 3.3, page 87) will help pupils make their models. They
should then compare the actual shapes of the models with the
displayed formulae. We often speak of straight chains of carbon
atoms. As pupils will see, these chains are far from straight!

The models they have made all have carbon atoms linked in
one continuous chain. Now is the time to introduce the idea
that, with alkanes containing four or more carbon atoms, the
chains may be branched. This leads to the existence of different
compounds with the same molecular formula but different
structures. These are called *isomers*. (Pupils often confuse this
word with 'isotopes' and 'allotropes'.)

Figure 3.4
The isomers of
C_4H_{10}.
a Butane.
b Methylpropane.

a

$$H-C-C-C-C-H$$

b

$$H-C-H$$

$$H-C-C-C-H$$

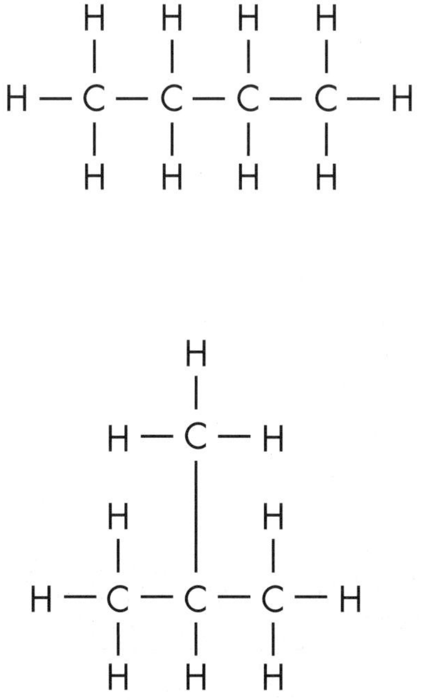

An example will make this clearer. There are two different
alkanes with the molecular formula C_4H_{10} – butane and
methylpropane (Figure 3.4). Pupils should make models of
butane and methylpropane. Then they could consider possible
isomers of pentane (Figure 3.5, overleaf). They should gain the
idea that the more carbon atoms in an alkane molecule, the
more isomers are possible.

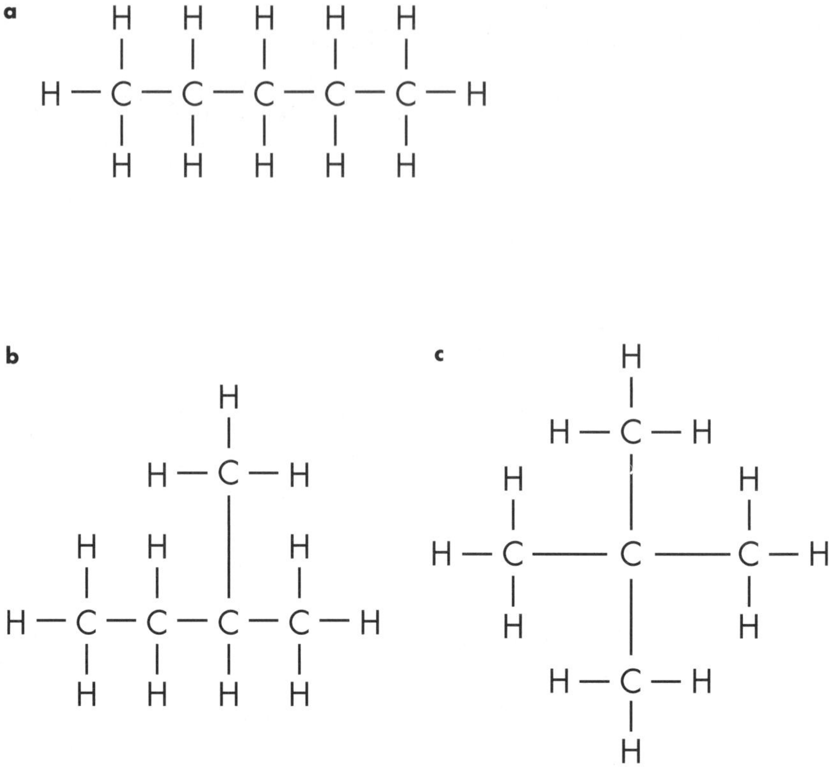

Figure 3.5
The isomers of
C$_5$H$_{12}$.
a *Pentane.*
b *Methylbutane.*
c *Dimethylpropane.*

The different shapes of the two isomers of C$_4$H$_{10}$ cause them to have different boiling points. The unbranched butane has a boiling point of $-1°C$, whereas the compact shape of methylpropane reduces the van der Waals forces between molecules, giving a lower boiling point of $-12\,°C$.

It is also possible to have alkanes in which the chains of carbon atoms are joined to make circular structures; these are called 'cyclic' molecules. An example is cyclohexane. Note that the molecular formula of straight-chain hexane is C$_6$H$_{14}$ but the molecular formula of cyclohexane is C$_6$H$_{12}$.

◆ *Enhancement ideas*

◆ There is much opportunity for a more detailed look at the effect of our use of alkanes on the environment. Petroleum products contain some sulphur, which adds to the problem of acid rain when the products are burned. Methane itself is a very powerful greenhouse gas.

3.4 Uses of alkanes
♦ *Previous knowledge and experience*

Pupils are likely to have come across (and used) alkanes such as methane and propane for cooking, although they may not know their names.

♦ *A teaching sequence*

The only common reactions of alkanes involve combustion.

Chemical properties of alkanes

The very strong covalent bonds between carbon and carbon atoms and between carbon and hydrogen atoms in alkanes means that a great deal of energy is needed to separate the atoms. In chemistry, the energy needed to separate two atoms which are bonded together is called the 'bond enthalpy'. It is measured in kilojoules of energy for each mole of bonds, kJ/mol. Some values of bond enthalpies are shown in Table 3.4.

Table 3.4 *Mean bond enthalpy values.*

Bond	Mean bond enthalpy (kJ/mol)
C—H	413
C=O	743
O=O	496
O—H	463
C—C	347

Any reaction involving an alkane will begin by breaking bonds, and will then go on to form new bonds in the product. The high bond enthalpies in alkanes mean that a lot of energy must be available for the initial breaking of bonds before a reaction can take place; in other words, reactions involving alkanes will have a high 'activation energy'. For this reason alkanes do not undergo many reactions.

Alkanes will undergo reactions where enough energy is supplied to provide the activation energy. The most important of these reactions, in terms of the quantity of alkanes used worldwide, is combustion. Most of the fractions containing smaller alkanes, e.g. paraffin, petrol and diesel, are burned as fuels.

The complete combustion of any alkane produces carbon dioxide and water, for example

methane + oxygen → carbon dioxide + water
CH_4 + $2O_2$ → CO_2 + $2H_2O$

Pupils are helped here by a graphical representation of the equation:

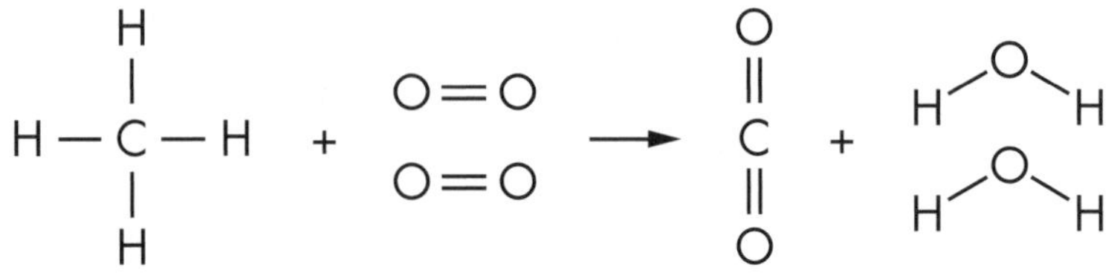

Pupils can work out the energy required to break the bonds within the reactants and the energy released when the bonds in the products are formed. For the energy required to break the bonds in reactants:

Bond broken	Number	Energy required (kJ)	
C—H	4	413 × 4	1652
O=O	2	496 × 2	992
		Total	2644

For the energy released when the bonds in the products are formed:

Bond formed	Number	Energy released (kJ)	
C=O	2	743 × 2	1486
O—H	4	463 × 4	1852
		Total	3338

The difference between the two totals is the overall energy released when 1 mole of methane is burned in excess oxygen. The energy change is −694 kJ/mol; the negative value, by convention, means that the reaction is exothermic.

Pupils can work out energy changes in other reactions, for example:

- the combustion of methane to produce carbon monoxide and water
- the combustion of ethane to produce carbon dioxide and water
- the combustion of methanol in excess oxygen.

In each case they will need to write a balanced equation and then to identify the number of each type of bond broken or formed. They could compare the values they obtain with data book values. These energy changes can be represented in an *energy diagram*, as in Figure 3.6.

Figure 3.6
Energy diagram for an exothermic reaction, e.g. combustion of an alkane.

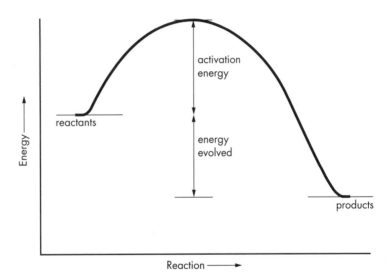

A mixture of an alkane, such as methane, and oxygen will not react at room temperature unless some energy is supplied, e.g. by a match or a spark. This supplies the activation energy for the first few molecules to react. When these molecules form products a large amount of energy is released, which in turn provides the activation energy for many more molecules to react. The reaction proceeds at a fast rate, releasing a large quantity of energy – hence the use of alkanes as fuels.

Carbon monoxide poisoning
If the supply of oxygen is restricted, not all of the atoms in the alkane will combine fully with oxygen. Carbon and carbon monoxide are produced. The carbon makes the flame sooty, and the carbon monoxide produced is very poisonous. Every

year there are fatal accidents involving carbon monoxide produced from faulty appliances such as gas fires, often in rented accommodation or holiday flats abroad. In 1998, 21 people died of carbon monoxide poisoning in the UK. If gas fires and water heaters are not regularly serviced or if ventilation is restricted, carbon monoxide is produced. This gas is particularly dangerous because it has no smell. You could show pupils examples of carbon monoxide detectors which can now be purchased.

Carbon monoxide is also released from car exhausts, though this problem has been improved in recent years by the introduction of catalytic converters, which catalyse the oxidation of carbon monoxide to carbon dioxide.

Heat output of different alkanes

This experiment can be carried out by pupils, or you can demonstrate it.

Materials
- eye protection
- copper calorimeter
- boss, stand and clamp
- thermometer, -10 to $110\,°C$
- butane lighter or similar
- measuring cylinder
- access to a balance
- mineral wool
- industrial methylated spirits

Safety
- *Butane is highly flammable.*
- *Ethanol is highly flammable. Industrial methylated spirit contains methanol, which is toxic.*
- *This experiment needs to be carefully controlled. The mixture of flames and flowing, highly flammable liquids can be a recipe for disaster.*
- *A demonstration on an open bench should be surrounded by safety screens.*
- *The use of the butane lighter must be carefully supervised as it can get hot with use.*
- *Use small bottles of the flammable liquids and locate them in one part of the lab, away from the flames.*
- *Absorb small amounts of the flammable liquids on to mineral wool.*
- *Wear eye protection when burning butane or other fuels.*

Procedure
1. Place $100\,cm^3$ of water in the calorimeter.
2. Find the mass of the butane lighter.
3. Measure the temperature of the water.
4. Light the butane lighter and hold it under the calorimeter (see Figure 3.7) until the water temperature has risen by a few degrees.
5. Re-weigh the butane gas lighter.

Figure 3.7
Measuring the heat output of an alkane.

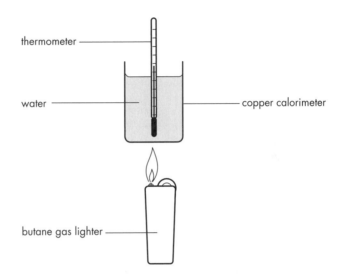

Since $1\,cm^3$ of water has a mass of $1\,g$ at room temperature, the heat output of the lighter, in joules, can be found from the experimental results by the following calculation:

heat output = mass of water (g) × temperature rise (°C) × 4.2

The heat output can then be divided by the mass of butane used to give the energy output per gram of the alkane.

The output can be compared with those of other fuels such as ethanol (methylated spirits) by using a similar method for each fuel; liquid fuels can be burned in a metal crucible. Use a small tuft of mineral wool to absorb the liquid and so avoid the risk of spillage.

The results will illustrate the large amount of heat energy released from a small mass of alkane, emphasising the use of alkanes as the main fuels of the second half of the 20th century.

The theoretical value for the energy given out when $1\,g$ of butane is burned in excess air is $49.8\,kJ$. The value obtained by pupils will be less than this, owing to the loss of energy to the surroundings.

Combustion products of a candle

In this experiment you can demonstrate the fact that carbon dioxide and water are produced when alkanes burn. The most convenient alkanes to burn are the large molecules which make up candle wax.

Materials
- eye protection
- candle
- funnel
- glass and rubber tubing
- U-tube in ice bath
- side-arm test-tube fitted with a delivery tube
- lime water
- suction pump (water-driven or hand)
- anhydrous copper(II) sulphate

Safety
- *Anhydrous copper(II) sulphate is harmful.*
- *The apparatus should be used behind safety screens.*
- *Wear eye protection when the candle is alight.*

Figure 3.8
Showing the combustion products of a candle.

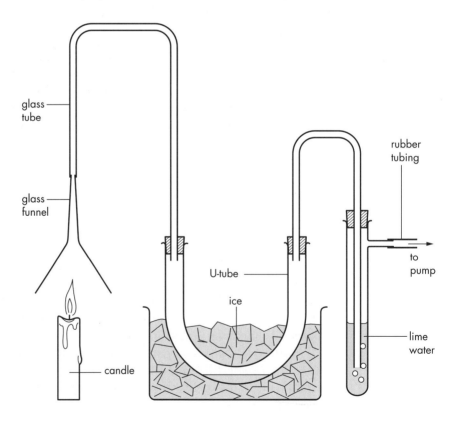

<u>Procedure</u>
1. Connect the apparatus as shown in Figure 3.8.
2. Turn on the suction pump.
3. Light the candle.
4. Observe the lime water over a few minutes.
5. After about 10 minutes, observe the liquid in the U-tube.

Note that water-driven suction pumps are not legal in some areas.

<u>What you might expect</u>
The lime water (a solution of calcium hydroxide) changes from a colourless, transparent liquid to a white suspension.

calcium hydroxide + carbon dioxide → calcium carbonate + water
$$Ca(OH)_2 \quad + \quad CO_2 \quad \rightarrow \quad CaCO_3 \quad + H_2O$$

The white precipitate is insoluble calcium carbonate. This shows the presence of carbon dioxide in the gases from the combustion.

Do not be surprised if, after a few minutes, the white precipitate disappears. As more carbon dioxide passes through, a further reaction takes place, producing calcium hydrogencarbonate, which is soluble (see Chapter 1):

calcium carbonate + water + carbon dioxide → calcium hydrogencarbonate
$$CaCO_3 \quad + H_2O + \quad CO_2 \quad \rightarrow \quad Ca(HCO_3)_2$$

The liquid in the U-tube may be tested with anhydrous copper sulphate, which changes from white to blue in the presence of water.

The funnel and glass tubing will become black owing to soot released from incomplete combustion of the candle. You can use this observation to emphasise what happens when alkanes burn in an insufficient supply of air. In this case, not enough oxygen can reach the candle flame, and so the alkanes burn with a luminous, sooty flame.

Pollution from alkanes

The use of alkanes as fuels is a major source of atmospheric pollution.

The most serious problem is caused by carbon dioxide. This is a 'greenhouse' gas, retaining heat energy from the sun in the atmosphere. The 'greenhouse effect' is vital for life on Earth as without it the average temperature would be about $-18\,°C$. However, as our use of alkanes as fuels (particularly for vehicle engines) has risen, so has the proportion of carbon dioxide in the atmosphere. This may be causing global warming. Many people think the greenhouse effect and global warming are the same thing, and it is important that pupils should learn the difference.

The overall rise in global temperature does not mean that all parts of the Earth's surface will get hotter. Changes in weather patterns will accompany global warming and will mean that some areas will get hotter and others colder, some wetter and others drier. Many areas are likely to suffer floods, while others may suffer droughts. In both cases, loss of crops may lead to famine.

Only strict control of the use of alkane fuels, particularly petrol, is likely to avert the catastrophic consequences predicted by many environmentalists. While some progress has been made, governments in general are reluctant to take measures which will have serious economic ramifications.

The carbon monoxide from car exhausts is also a problem. Although it will combine with oxygen in the air to form carbon dioxide, this process is slow in urban areas. In city centres the carbon monoxide content of the air rises as traffic density increases. When carbon monoxide is inhaled, it combines irreversibly with the haemoglobin in red blood cells; this reduces the capacity of the blood to carry oxygen. People living or working in cities may have a reduced ability to transport oxygen, similar to those who smoke cigarettes, which also produce carbon monoxide.

Pupils could collect information on the greenhouse effect and global warming from newspapers, magazines, CD ROMs (e.g. *Encarta*), etc.

◆ *Enhancement ideas*

- ◆ Pupils could study the combustion of the homologous series of alcohols. Small spirit lamps containing methanol, ethanol, propanol, butanol and pentanol can be used to heat calorimeters containing 100 cm^3 of water. The temperature rise is measured each time. The mass of the spirit lamp before and after use will give the mass of alcohol used. Pupils can calculate the energy released when one mole of each alcohol is burned. They can use average bond enthalpies to calculate theoretical values. This experiment is a very good activity for pupils aiming at high marks in their science investigation skills.

3.5 Cracking alkanes to produce alkenes

♦ *Previous knowledge and experience*

During their study of the fractional distillation of crude oil, pupils may become aware that there is not a perfect match between the percentages of the different fractions in the crude oil and the demands of consumers. They should be anticipating that any process which helps to use the higher boiling point fractions to produce useful products will be economically beneficial.

♦ *A teaching sequence*

Catalytic cracking

The proportions of different fractions in crude oil vary according to the location of the oil wells (see Table 3.2, page 85). In general, however, the percentage of crude oil distilled into the petroleum fraction is far less than the percentage usage of this fraction. To obtain more of the alkanes of the size found in petrol it is possible to split up larger alkanes. This reaction is known as 'cracking'. For example:

decane $\rightarrow$ octane $+$ ethene

$C_{10}H_{22}$ $\rightarrow$ C_8H_{18} $+$ C_2H_4

The reaction will take place at high temperatures, but is normally carried out using a combination of a moderately high temperature and a catalyst such as aluminium oxide. A frequent mistake made by pupils is to write that a high *pressure* is needed; this is not the case.

From a range of larger alkanes, the smaller alkanes which comprise the petrol fraction are produced. Thus a greater proportion of the petrol fraction is obtained from the crude oil, to satisfy the growing demand for vehicle fuel.

The other product of catalytic cracking is another type of hydrocarbon, an alkene. In the example shown above the alkene produced is ethene, though other alkenes are also produced in the cracking process.

Alkenes are another homologous series, characterised by the presence of a double covalent bond between two carbon atoms (see Figure 3.9).

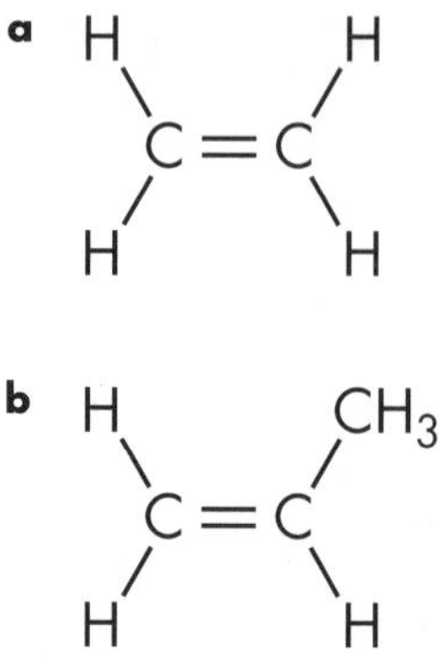

Figure 3.9
The first two alkenes.
a *Ethene.*
b *Propene.*

Alkenes have the general formula C_nH_{2n}. The presence of a double covalent bond between two carbon atoms makes alkenes far more reactive than alkanes. They are of immense importance in chemical manufacturing, as the raw materials from which many useful chemical compounds are made.

Pupils can make a long alkane chain from 'poppet' beads or paperclips. They can then break up the chain into short lengths of one or two beads (or clips) to demonstrate the process of cracking. When you move on to study polymerisation, this can be considered as the opposite of cracking.

Cracking alkanes

Pupils can perform this as a class experiment, or you may prefer to demonstrate it.

The catalyst can be granular aluminium oxide (or, more cheaply, some broken pottery which contains aluminium oxide). The mineral wool soaks up the hydrocarbon, preventing it from running out of the horizontal tube. A suitable hydrocarbon mixture to use is medicinal paraffin.

<u>Materials</u>
- eye protection
- boiling tube
- aluminium oxide
- mineral wool
- glass delivery tube
- trough
- boss, stand and clamp
- several test-tubes with corks to fit
- test-tube rack
- Bunsen burner
- heatproof mat
- medicinal paraffin
- bromine water

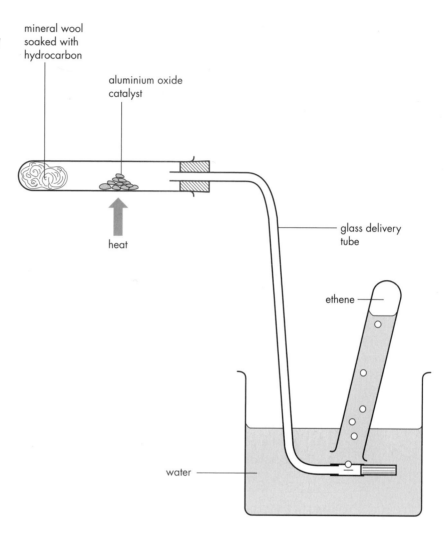

Figure 3.10
Catalytic cracking of an alkane.

mineral wool soaked with hydrocarbon

aluminium oxide catalyst

heat

glass delivery tube

ethene

water

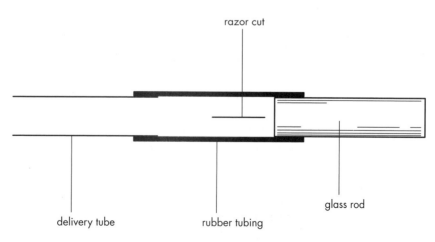

Figure 3.11
A Bunsen valve.

razor cut

glass rod

delivery tube

rubber tubing

<u>Safety</u>
- *1% bromine water (0.06 mol/dm³) is toxic and corrosive; 0.1% bromine water (0.006 mol/dm³) is harmful and irritant. Use a very dilute solution to avoid problems.*
- *Care must be taken to remove the apparatus from the trough when heating is discontinued, to prevent 'suck-back' of the water. This occurs when heating of the boiling tube is reduced: the gas inside the apparatus contracts and water enters the delivery tube to fill the space. If sucking back starts, heat the boiling tube more strongly, or, if the paraffin is almost used up, lift the delivery tube out of the water.*
- *A demonstration on an open bench should be surrounded by safety screens.*
- *Wear eye protection when using a Bunsen burner.*

<u>Procedure</u>
1. Set up the apparatus as shown in Figure 3.10. The end of the delivery tube should have a 'Bunsen' valve attached; this is simply a short piece of rubber tubing which has been sliced along its length with a razor blade and then blocked at the end with a length of glass rod (see Figure 3.11). It allows gas to escape but prevents water entering the tube and being sucked back on to hot glass.
2. Heat the aluminium oxide strongly.
3. Collect the gas given off in test-tubes over water.
4. Try burning the gas in one test-tube.
5. To the gas in another test-tube, add a few drops of bromine water and shake.

<u>What you might expect</u>
During the cracking, droplets of the shorter alkane will be seen floating on the water in the trough. The gaseous alkene, ethene, will be collected in the test-tubes, which should be corked under water and placed in a test-tube rack ready for testing.

When a lighted splint is applied to the gas in one tube, a flame should be seen moving slowly along the tube. Like other alkenes, ethene is flammable. Alkenes generally burn with a smokier flame than alkanes because of their higher percentage of carbon. They are not normally used as fuels since they are far more useful in the chemical industry in the production of a wide range of materials.

When bromine water is shaken with the gas, the orange colour of the bromine disappears. Bromine water can be used as a test to distinguish alkanes from alkenes: the former will not decolorise bromine. The orange colour of bromine disappears because the bromine molecules are reacting with the alkene. In the example of ethene, the reaction is:

The product is 1,2-dibromoethane, which is colourless. Since the bromine has been added to the alkene molecule, this type of reaction is known as an 'addition reaction'.

Pupils are not expected to know the details of industrial cracking processes, which are kept as 'secrets' by the oil companies.

◆ *Enhancement ideas*

◆ A similar apparatus to the one used in the cracking of medicinal paraffin can be used for the dehydration of ethanol:

$$C_2H_5OH \rightarrow C_2H_4 + H_2O$$

Bromine water can be used to test for the ethene produced. (Note: ethanol is highly flammable; industrial methylated spirit contains methanol, which is toxic; 1% bromine water (0.06 mol/dm^3) is toxic and corrosive, 0.1% bromine water (0.006 mol/dm^3) is harmful and irritant.)

◆ Addition reactions to ethene can produce useful products. Pupils could construct models or displayed formulae of the products of adding hydrogen, water and hydrogen bromide to ethene.

◆ *Further activities*

◆ *Classic Chemistry Demonstrations* contains the following demonstrations which are relevant to this section:
    50   Unsaturated compounds in food
    60   The reaction of ethyne with chlorine
    98   Cracking a hydrocarbon/dehydrating ethanol

◆ *Microscale Chemistry* contains the following experiments which are relevant to this section:
    50   Testing for unsaturation using potassium manganate(VII)
    51   Preparing and testing ethyne
    52   Testing for unsaturation using bromine

3.6 Polymers

◆ *Previous knowledge and experience*

Pupils will be familiar with some polymers from primary school, but will probably call them 'plastics'. Poly(ethene), called by its traditional name polythene, is used for packaging and for everyday objects. Poly(phenylethene), called polystyrene, is used for ceiling tiles and plastic models. Pupils will know some of the useful properties of plastics.

◆ *A teaching sequence*

Using models

One of the important reactions of alkenes carried out in the chemical industry is *addition polymerisation*. In this reaction molecules of alkene, called *monomers*, add on to each other to form a long chain, called the *polymer*. The best way of introducing what is happening is to have a pile of unconnected 'poppet' beads and then join them together to form a long chain. (Paperclips can be used instead of 'poppet' beads.)

Figure 3.12 shows the formation of a small part of the polymer from ethene – poly(ethene). Pupils can use molecular model kits to construct several models of the monomer, and then 'polymerise' this into a model of the polymer.

Figure 3.12
Addition polymerisation of ethene to form poly(ethene) (polythene).

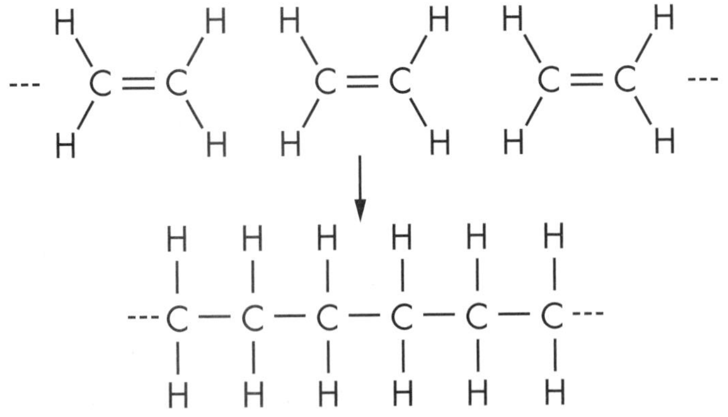

Polymers can be made from many other compounds containing a carbon–carbon double bond; some of these are alkenes, while others are not hydrocarbons since they contain atoms of other elements as well as hydrogen and carbon. Some examples are shown in Table 3.5.

Table 3.5 *Some polymers formed by addition polymerisation.*

Monomer		Polymer	
Ethene	$CH_2{=}CH_2$	Poly(ethene) (polythene)	$-(-CH_2-CH_2-)_n$
Propene	$CH_2{=}CHCH_3$	Poly(propene)	$-(-CH_2-CHCH_3-)_n$
Phenylethene (styrene)	$CH_2{=}CHC_6H_5$	Poly(phenylethene) (polystyrene)	$-(-CH_2-CHC_6H_5-)_n$
Chloroethene (vinyl chloride)	$CH_2{=}CHCl$	Poly(chloroethene) (PVC)	$-(-CH_2-CHCl-)_n$
Tetrafluoroethene	$CF_2{=}CF_2$	Poly(tetrafluoroethene) (PTFE or Teflon)	$-(-CF_2-CF_2-)_n$

When small monomer molecules combine, a double bond in each molecule is broken and a single bond remains.

Each polymer has different properties which make it suitable for specific purposes; for example, PTFE has non-stick properties and is used to coat frying pans and saucepans.

◆ *Enhancement ideas*

- ◆ Pupils could set up an exhibition of objects made from polymers.
- ◆ A study of polymers could be extended to include condensation polymerisation.
- ◆ Nylon could be made as a class demonstration (see No. 64 in *Classic Chemistry Demonstrations*, mentioned below).
- ◆ Using paperclip chains to represent polymers can be extended to consider links between chains. These cross-links produce polymers which do not melt easily. Pupils could study thermosetting and thermoplastic polymers.

◆ *Further activities*

- ◆ *Classic Chemistry Demonstrations* contains the following demonstrations which are relevant to this section:
    - 15 Urea–methanol polymerisation
    - 21 Phenol–methanol polymerisation
    - 31 Disappearing plastic
    - 64 Making nylon – the 'nylon rope trick'
    - 91 Making rayon
    - 95 Making polysulphide rubber

♦ *References*

Classic Chemistry Demonstrations (1998). Royal Society of Chemistry. ISBN 1 870343 28 7. This book was distributed free to all schools in the UK. Further information is available from Royal Society of Chemistry, Burlington House, Piccadilly, London W1V 0BN.

Microscale Chemistry (1998). Royal Society of Chemistry. ISBN 1 870343 49 2. Many experiments that cannot be carried out easily on a large scale (except in a fume cupboard), because of the toxicity of reactants or products, can be carried out on a very small scale in the open lab. This book gives many possibilities, and is circulated free to all secondary schools in the UK. Further information from the RSC.

♦ *Other resources*

A useful book on carbon chemistry is *Plastics a Plenty* (1990) by Bob McDuell (ISBN 0 7487 0210 5) published by Stanley Thornes, Old Station Drive, Leckhampton, Cheltenham GL53 0DN. This includes a key which can be used to identify some common polymers.

 The video *Industrial Chemistry for Schools and Colleges* produced by the RSC includes information about industrial processes. (For further details, see Chapter 9.)

Much of the work in carbon chemistry is enhanced by the use of models. Simple models can be made with Plasticine balls and cocktail sticks. A wide range of other models can be purchased from Cochranes of Oxford Ltd, Leafield, Witney OX8 5NY.

A range of materials is available from the Institute of Petroleum, 61 New Cavendish Street, London W1M 8AR. These include a series of oil data sheets and booklets for secondary school children, including *Story of Oil*, *Oil and Gas: Energy for the World* and *Fossils into Fuels*.

Educational materials on carbon chemistry are available from Resources Plus Ltd–ESSO at Ringway House, Kelvin Road, Newbury, Berkshire RG14 2OB.

 Web sites

♦ **www.chevron.com/explore/main.html** provides data on the formation, exploration and refining of crude oil.

♦ Poly(chloroethene) (polyvinyl chloride) is a widely used polymer. Its advantages and disadvantages are discussed on: **www.bpf.co.uk/options.htm**

4 Extracting metals from rocks

Neil Rowbotham

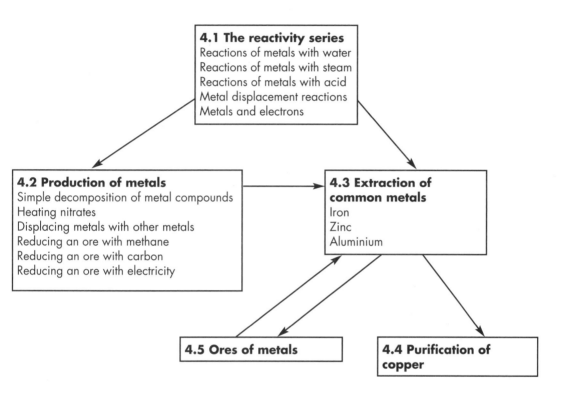

4.1 The reactivity series
Reactions of metals with water
Reactions of metals with steam
Reactions of metals with acid
Metal displacement reactions
Metals and electrons

4.2 Production of metals
Simple decomposition of metal compounds
Heating nitrates
Displacing metals with other metals
Reducing an ore with methane
Reducing an ore with carbon
Reducing an ore with electricity

4.3 Extraction of common metals
Iron
Zinc
Aluminium

4.5 Ores of metals

4.4 Purification of copper

◆ Choosing a route

The method used for producing a particular metal is not chosen on a whim – there are fundamental principles behind the choice and they should be introduced in a straightforward manner, without over-elaboration but without ignoring some aspects. The main consideration is the *reactivity* of the metal. It is worth spending time showing pupils that the more reactive the metal is, then the greater will be the difficulty in producing it from its ores. Begin with reactivity; this should then lead to investigating how the reduction of the ore might be carried out in practice. Alternative routes after that could be to look at specific examples or to investigate the available ores and see if they could be converted to the metal by a variety of methods.

4.1 The reactivity series

◆ *Previous knowledge and experience*

Pupils will have little experience of chemical reactions from primary school science.

◆ *A teaching sequence*

The concept of reactivity is not an easy one to explain in detail but it would be sufficient to begin with the idea that 'the faster a metal reacts, then the more reactive it is'.

Reactions of metals with water

The starting point might be a series of teacher demonstrations showing the reactivity of a series of metals with water. One might choose to show them in the accepted order or to exhibit their behaviour at random.

Demonstration of the reactions of sodium, potassium and lithium with water

Materials
- eye protection
- safety screens
- large trough (glass or clear plastic)
- sheet of glass or plastic to act as a lid
- pair of tongs
- knife
- Bunsen burner
- wooden splints
- lithium
- potassium
- sodium
- universal indicator solution

Safety
- *Lithium, potassium and sodium are flammable and very corrosive.*
- *Only the teacher should carry out experiments with sodium, potassium and lithium.*
- *The metals should be picked up with tongs and never handled.*
- *Always use clean, fresh samples of these metals.*
- *Do not add universal indicator to the water until all reaction has ceased.*
- *Pupils should be 2–3 m from the experiments.*
- *The teacher must use safety screens and must wear eye protection.*

Procedure

1. Half-fill the trough with tap water.

2. Using a pair of tongs, remove a sample of sodium from the bottle. Show how it is kept under oil to keep it away from air and water, and lead the class to appreciate that this suggests potential reactivity. Holding the sodium with the tongs, cut a series of pieces about 0.25 cm cubed. As you do this, explain to the class how easy sodium is to cut. (This is very unusual for a metal – perhaps link to metals they know and the difficulty they have in imagining cutting them with a small knife.)

3. Show how shiny the surface of the metal is to convince them that sodium *is* a metal despite its softness.

4. Drop one piece of sodium on to the surface of the water in the trough and place the lid on top of the trough at once. Help the pupils to notice that the sodium reacts so vigorously with the water that it melts with the heat produced.

Note how the molten silver ball of sodium moves across the surface of the water and lead the pupils to consider a 'hovercraft' effect, where the molten sodium is actually moving on a layer of gas. Apply a lighted splint to the gas above the moving sodium and show that it burns and that the flame is coloured yellow by the presence of the sodium metal.

To do all that will probably take five or six pieces of sodium added to the water one at a time to allow you to bring out each point. Be warned – the water could get hot, leading to a too violent reaction. If necessary, replace the water. Replace the lid after every addition. Resist the temptation to add lots of pieces to the water at once, and firmly resist calls to add bigger pieces!

5. When all the sodium has disappeared, add universal indicator to the water. It will turn blue, showing that an alkali has been produced which has dissolved in the water – a strong alkali at that, since very little can have been made and yet it has a distinct effect on the indicator.

A summary word equation can be developed from:

sodium + water → alkali + flammable gas

leading to

sodium + water → sodium hydroxide + hydrogen

The symbol equation is

$$2Na + 2H_2O \rightarrow 2NaOH + H_2$$

6. Now repeat the experiment with potassium but:

- make sure that all the sides of a 0.25 cm cube of potassium are cleanly cut and shiny before you put it into the water
- before adding the potassium, cover nearly all the trough with the lid
- do not attempt to apply a lighted splint to the moving potassium.

In this case the potassium will melt to a silver ball with the heat of the reaction and will move across the surface of the water, still producing enough extra energy to set fire to the gas above it. The colour of the flame will be lilac because of the presence of the potassium. It is quite likely that as each piece of potassium finishes reacting there will be a very small crack as the reaction reaches explosive conditions. Pupils must be well protected by the lid and safety screen and at a reasonable distance.

Adding universal indicator to the water remaining at the end shows that an alkali has again been formed as one of the products of the reaction:

$$\text{potassium} + \text{water} \rightarrow \text{alkali} + \text{flammable gas}$$
$$2K + 2H_2O \rightarrow 2KOH + H_2$$

7. The experiment may be repeated with lithium, which behaves in a similar fashion except that:

- it is harder to cut than sodium, and much harder than potassium
- it sinks in the water at first, before being lifted to the surface by bubbles of the gas being produced
- it does not produce enough heat to melt the sample
- the gas given off does not self-ignite but it can be ignited by a burning splint and is coloured red by the presence of the lithium.

If you are unhappy showing these demonstrations to a particular class, then they can be seen on the video *Classic Chemical Demonstrations* (Department of Chemistry, University of Leeds). Alternatively, the CD ROM *The Elements* has video clips of these reactions.

Reactions of calcium, magnesium and aluminium with water

Materials
- eye protection
- large trough (glass or clear plastic)
- sheet of glass or plastic to act as a lid
- pair of tongs
- knife
- Bunsen burner
- wooden splints
- beaker
- filter funnel
- test-tube
- calcium
- magnesium
- aluminium
- universal indicator solution

Safety
- *Calcium and magnesium are highly flammable.*
- *In contact with moisture calcium forms calcium hydroxide, which is an irritant to eyes and skin.*
- *Wear eye protection when using calcium and magnesium.*

Procedure
1. Show that calcium is not kept under oil and help the class to suggest that this might mean it is less reactive than those metals considered so far.
2. Half-fill the trough with tap water and add one or two pieces of calcium metal. The calcium will sink and begin reacting steadily. Bubbles of gas will be given off and these will probably lift the calcium to the surface, but it will fall when the gas is released into the atmosphere. The calcium will not melt.
3. Adding five or six pieces of calcium will produce quite a lot of gas and it is possible to chase the bubbles around the surface of the water with a lighted splint and ignite the bubbles with a 'pop' and a flash of red colour. Pupils will quickly suggest that the gas is hydrogen, if prompted.
4. As the reaction proceeds, the water will become cloudy with a white solid. This could be linked to solubility and a suggestion that the product of this reaction does not dissolve as easily in water as those investigated in previous experiments.
5. When all reaction has ceased, add a little universal indicator to show that an alkali has been produced (the indicator will turn blue):

$$Ca + 2H_2O \rightarrow Ca(OH)_2 + H_2$$

6. Pupils can then repeat the above experiment with pieces of magnesium in a beaker of water: little will appear to happen.

7. Put a filter funnel over the top of the pieces of magnesium and submerge the whole apparatus under water in a sink. While still under water, place a test-tube, full of water, over the open end of the funnel then remove the whole set of apparatus from the sink. It should appear as in Figure 4.1. Leave until the next lesson.

Figure 4.1
Reacting magnesium ribbon with water.

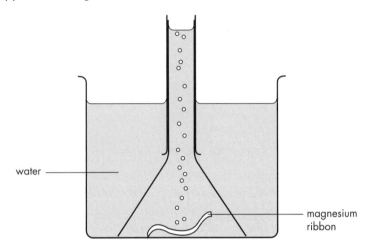

water ——————————

magnesium ribbon

Inspection then will show that a significant amount of gas has been produced by the reaction between the magnesium and the water, and that the gas has collected in the test-tube. If there is sufficient you could test it to show that it is hydrogen.

8. Repeat with aluminium.

All this work should lead the class to appreciate that potassium is the most reactive metal, sodium the next, lithium next, then calcium and magnesium, and finally aluminium. However, this applies only to the metals' reaction with water. To show their reactivity with oxygen or chlorine consult *Classic Chemistry Demonstrations* No. 72 (Royal Society for Chemistry, 1998, ISBN 1 870 343 38 7).

Reactions of metals with steam

Having established a reactivity series with water, it is important to go on to show that the same order may be obtained in other ways. As a class practical pupils may heat metals in steam using the apparatus in Figure 4.2.

The metals to use are zinc, magnesium, iron and copper. If you wish, calcium's reaction with steam may be shown as a teacher demonstration but do *not* attempt to react lithium, sodium or potassium with steam.

Materials
- eye protection
- hard glass test-tube
- delivery tube and bung (see Figure 4.2)
- beaker or trough of water
- test-tubes
- Bunsen burner
- heatproof mat, stand, boss and clamp
- mineral wool
- magnesium ribbon
- zinc powder
- iron powder
- copper powder

Figure 4.2
Reacting magnesium ribbon with steam.

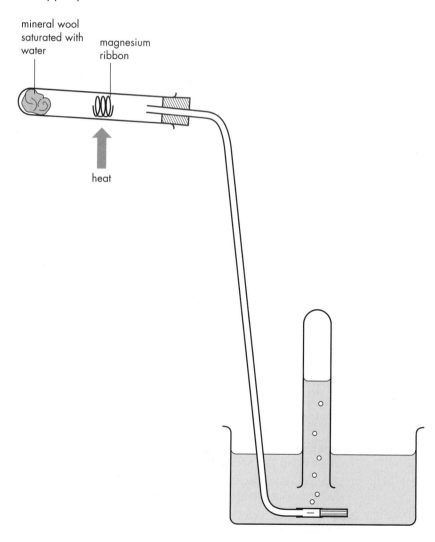

Safety
- *Magnesium is highly flammable.*
- *Magnesium will burn vigorously and is likely to damage the test-tube; it may cause it to break. After use, place the test-tube on a heatproof mat to cool.*
- *Moist zinc dust can ignite spontaneously in air.*
- *Wear eye protection when using a Bunsen burner.*

Procedure
1. Place about 2 cm depth of mineral wool in the bottom of a test-tube, fill the test-tube with water and allow it to stand while the rest of the apparatus is collected. (This allows the mineral wool to become saturated with water.)
2. Pour off all the excess water and, holding the test-tube horizontally, use a spatula to put a pile of metal half-way down the tube.
3. Still holding the test-tube horizontally, connect up the apparatus as in Figure 4.2. The end of the delivery tube should have a 'Bunsen' valve attached; refer back to Chapter 3, page 102, for details of how to set up a Bunsen valve.

   The test-tube containing the metal should be horizontal or slope very slightly towards the rubber bung, to ensure that any condensed water does not run on to the red-hot glass, causing the glass to crack.
4. Heat the *metal* with a Bunsen burner. Do not heat the mineral wool: heat will conduct down towards the mineral wool and evaporate water from it at a steady rate, whereas heating the wool itself will dry it out too quickly.
5. If any reaction occurs then a gas will be given off which may be collected over water. Testing with a burning splint should show that it is hydrogen. Hydrogen itself burns quietly with a blue flame, but the presence of even a small amount of air makes it burn with a squeaky pop.

What you might expect

Table 4.1 *Reactions of some metals with steam.*

Metal	Observations	Equation
Magnesium	Magnesium burns with a bright 'white' flame and turns to a white powder. Hydrogen gas given off in large quantities	$Mg + 2H_2O \rightarrow MgO + H_2$
Zinc	Zinc glows white-hot and turns to a powder which is yellow when hot and white when cold. A few test-tubes of hydrogen gas produced	$Zn + 2H_2O \rightarrow ZnO + H_2$
Iron	Slight change in colour as reaction proceeds but not much else to see. Some gas given off but may be less than one test-tube full	Reaction is reversible: $3Fe + 4H_2O \rightleftharpoons Fe_3O_4 + 4H_2$
Copper	No visible reaction	—

From these experiments we have

Mg > Zn > Fe > Cu

so overall

K > Na > Li > Ca > Mg > Zn > Al > Fe > Cu

Reactions of metals with acid

Do not attempt this with lithium, sodium or potassium. Calcium will react well with dilute hydrochloric acid, but in sulphuric acid it will quickly become coated with insoluble calcium sulphate and this will prevent it reacting further. Do not use dilute nitric acid since, as well as being an acid, it is a powerful oxidising agent and this can confuse the issue. Also, 1 mol/dm^3 nitric acid is corrosive.

Materials
- eye protection
- test-tubes
- delivery tube and bung
- large beaker
- splints
- pieces of magnesium, aluminium, zinc, iron and copper
- dilute hydrochloric acid (1 mol/dm^3)
- dilute sulphuric acid (1 mol/dm^3)

Safety
- *Sulphuric acid and hydrochloric acid are irritants, even when dilute, and should be kept off skin.*
- *Wear eye protection when using these acids. When the bubbles of gas burst, a spray of acid droplets will be produced.*

Procedure
1. Place two or three small pieces of metal or a spatula-end of metal powder in the bottom of a test-tube. Add 1 cm depth of dilute hydrochloric acid or sulphuric acid. Observe.
2. If a gas is seen to be produced try to collect it over water. The gas should be hydrogen.

Table 4.2 *Reactions of some metals with acids.*

Metal	Reaction with hydrochloric acid	Reaction with sulphuric acid
Magnesium	Fast reaction even with ribbon. Hydrogen produced: $Mg + 2HCl \rightarrow MgCl_2 + H_2$	Fast reaction even with ribbon. Hydrogen produced: $Mg + H_2SO_4 \rightarrow MgSO_4 + H_2$
Zinc	Quite fast with granules once it has begun, much faster with powder. Hydrogen produced: $Zn + 2HCl \rightarrow ZnCl_2 + H_2$	Quite fast with granules once it has begun, much faster with powder. Hydrogen produced: $Zn + H_2SO_4 \rightarrow ZnSO_4 + H_2$
Aluminium	Quite fast with powder but may take a while to begin. Hydrogen produced: $2Al + 6HCl \rightarrow 2AlCl_3 + 3H_2$	Quite fast with powder but may take a while to begin. Hydrogen produced: $2Al + 3H_2SO_4 \rightarrow Al_2(SO_4)_3 + 3H_2$
Iron	Steady reaction with powder. Hydrogen produced: $Fe + 2HCl \rightarrow FeCl_2 + H_2$	Steady reaction with powder. Hydrogen produced: $Fe + H_2SO_4 \rightarrow FeSO_4 + H_2$
Copper	No visible reaction	No visible reaction

The order of reactivity is

$Ca > Mg > Zn > Al > Fe > Cu$

Metal displacement reactions

Metals can be placed in order of their reactivity by 'playing each other' in a Metals' Premier League competition.

Materials
- eye protection
- white tile
- dropping pipettes
- pieces of calcium, magnesium, aluminium, zinc, iron and copper
- solutions of calcium nitrate, magnesium sulphate, aluminium sulphate, zinc sulphate, iron(II) sulphate and copper(II) sulphate (all about 1 mol/dm³; iron(II) sulphate solution should be freshly prepared from the solid as required)
- silver nitrate solution (about 0.05 mol/dm³)

Safety
- *Calcium nitrate is an oxidising agent, magnesium salts are of a low hazard, aluminium sulphate solution is of a low hazard, zinc sulphate is only hazardous if ingested in large quantities, iron salts and copper salts are harmful if swallowed and silver nitrate should be used as 0.05 mol/dm³.*
- *Since only drops are used the solutions should be safe unless misused.*
- *Wear eye protection when using silver nitrate solution and keep it off the skin.*

Figure 4.3
Finding the reactivity order in the Metals' Premier League.

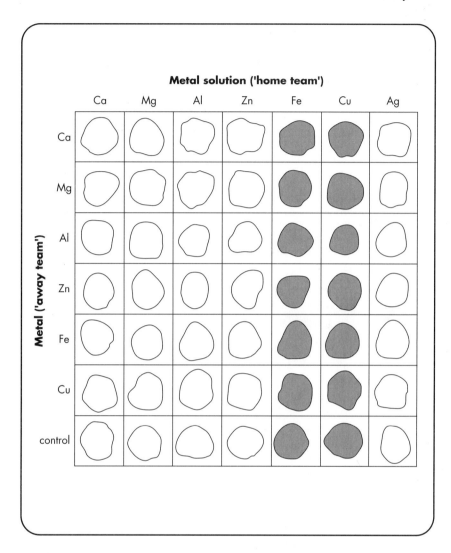

Procedure

1. On a white tile place drops of each of the seven solutions in columns in a 7 × 7 grid as shown in Figure 4.3: these are the 'home teams'.
2. Now add a very small piece of metal (the 'away team') to each of these drops and note any change. Any alteration at all is a 'win' for the visitor. Note that one row of drops does not have any metal added: this is a control and can be used for comparing.
3. Draw a league table and add up the points – three for a win and one for a draw.
4. Silver may be added to the league without the need to use expensive metallic silver. If you add the six metals to silver nitrate solution the metal will displace silver from solution every time. So silver must be the weakest team in the league.

What you might expect

Table 4.3   *The results of the 'league matches'.*

Metal ('away team')	Metal solution ('home team')						
	Calcium nitrate	Magnesium sulphate	Aluminium sulphate	Zinc sulphate	Iron(II) sulphate	Copper(II) sulphate	Silver nitrate
Ca	—	Away	Away	Away	Away	Away	Away
Mg	Home	—	Away	Away	Away	Away	Away
Al	Home	Home	—	Away	Away	Away	Away
Zn	Home	Home	Home	—	Away	Away	Away
Fe	Home	Home	Home	Home	—	Away	Away
Cu	Home	Home	Home	Home	Home	—	Away

The order of reactivity is

Ca > Mg > Al > Zn > Fe > Cu > Ag

Metals and electrons

All of these changes can be summarised as the metal losing electrons (*oxidation*). Typical ionic equations are:

$$K \rightarrow K^+ + e^-$$
$$Na \rightarrow Na^+ + e^-$$
$$Li \rightarrow Li^+ + e^-$$
$$Ca \rightarrow Ca^{2+} + 2e^-$$
$$Mg \rightarrow Mg^{2+} + 2e^-$$
$$Al \rightarrow Al^{3+} + 3e^-$$
$$Zn \rightarrow Zn^{2+} + 2e^-$$
$$Fe \rightarrow Fe^{2+} + 2e^-$$
$$Pb \rightarrow Pb^{2+} + 2e^-$$
$$Cu \rightarrow Cu^{2+} + 2e^-$$
$$Ag \rightarrow Ag^+ + e^-$$

Put in this sequence, the list of metals is known as the *reactivity series of metals.* The higher up the series a metal is, the more reactive it is, i.e. the more it will do and the faster it will do it. Any metal in the series will displace a metal below it from its salts, for example calcium will displace magnesium but zinc will not.

We often include hydrogen in the series even though it is not a metal. This is done to show that metals above hydrogen in the series will displace hydrogen gas when the metal is placed in acid; metals below it never produce hydrogen gas with acids. Metals below hydrogen will be produced when their oxides are heated and have hydrogen passed over them.

Table 4.4 *The reactivity series of metals – a summary.*

Metal	Reaction of metal with water	Reaction of metal with acid	Reaction of oxide with hydrogen	Other metals it displaces from their compounds
Potassium, K	Violent in cold water	Dangerous	No reaction	Na to Ag
Sodium, Na	Vigorous in cold water	Dangerous	No reaction	Li to Ag
Lithium, Li	Quick in cold water	Dangerous	No reaction	Ca to Ag
Calcium, Ca	Steady in cold water	Violent, giving hydrogen	No reaction	Mg to Ag
Magnesium, Mg	Slow in cold water, fast in steam	Very fast, giving hydrogen	No reaction	Al to Ag
Aluminium, Al	Slow	Slow to start but then fast, giving hydrogen	No reaction	Zn to Ag
Zinc, Zn	No reaction with cold water, steady in steam	Steadily, giving hydrogen	No reaction	Fe to Ag
Iron, Fe	No reaction with cold water, reversible in steam	Slowly, giving hydrogen	Some iron produced, but very slowly	Pb to Ag
Lead, Pb	No reaction	Reacts slowly with concentrated acid	Lead produced very slowly	Cu to Ag
Hydrogen, H	—	—	—	Cu to Ag
Copper, Cu	No reaction	No reaction	Copper produced quickly	Ag
Silver, Ag	No reaction	No reaction	Silver produced quickly	None

♦ *Enhancement ideas*

- ♦ Pupils could study the topic of corrosion of metals. Rusting of iron and steel in Great Britain costs £2 000 000 000 each year, and methods of reducing rusting are big business. Simple rusting experiments with nails in different conditions will show that water and oxygen are needed for rusting to take place. Interesting experiments with ferroxyl indicator (a mixture of potassium hexacyanoferrate(III) and phenolphthalein solutions) can show where rusting takes place.
- ♦ Pupils could be given another metal not in their reactivity series. They could then plan and carry out experiments to find where the metal will be placed in the reactivity series. Suitable metals are manganese or nickel. Note that nickel is harmful and may cause sensitisation by skin contact.

4.2 Production of metals

♦ *Previous knowledge and experience*

Pupils are likely to have come across ideas about the extraction of metals in History and Geography, for example, iron in a study of the Iron Age.

♦ *A teaching sequence*

It may take time to convince the class that the production of a metal from its ore involves the very reverse of the processes they have studied and that the more reactive the metal, i.e. the more easily it loses electrons to form ions, then the harder it will be to reverse that process. Adding electrons is called *reduction*.

A useful analogy here is to consider water running over a waterfall. If the water falls only a few centimetres it makes very little noise, has little power and can easily be splashed back over the stone. If the water drops a considerable distance then it makes a lot of noise, hits the ground with power and a lot of energy is required to pump it back up again. Reactive metals produce a lot of energy when they react and require a lot of energy to extract them. Unreactive metals produce little energy when they react and require little energy to extract them.

Simple decomposition of metal compounds: carbonates, oxides and sulphates

Heating compounds is a good way to decompose them. Pupils can try to see if it is possible to obtain the metals in this way. Compounds of metals which are commonly found in rocks are carbonates, sulphates and oxides.

Pupils will quickly discover that simple heating is a very unproductive approach, so it is unwise to ask them to attempt all the compounds – far better to have them do one or two each, with some cross-checking between the groups, and then pool the results.

<u>Materials</u>
- eye protection
- test-tubes
- test-tube holder
- Bunsen burner
- spatula
- heatproof mat
- samples of copper(II) carbonate, copper(II) oxide, lead(II) carbonate, zinc carbonate, magnesium carbonate, calcium carbonate and zinc oxide
- dilute hydrochloric acid (1 mol/dm³)

<u>Safety</u>

- *Lead compounds are toxic.*
- *Copper compounds are harmful.*
- *Heating magnesium carbonate produces the oxide, which is a mild alkali when wet.*
- *Heating calcium carbonate produces the oxide, which is a corrosive alkali when wet.*
- *Hydrochloric acid is an irritant, even when dilute, and should be kept off skin.*
- *Keep all dust to an absolute minimum and wipe up with a damp cloth.*
- *Wear eye protection.*

<u>Procedure</u>
1. Place enough of a compound to cover a spatula end in a test-tube, and heat it using a hot blue Bunsen flame.
2. Cool, by placing the test-tube on the heatproof mat. When cool, inspect the product to see if it shows any metallic appearance or if hydrogen is produced with dilute hydrochloric acid (see page 115).

Heating nitrates

Although nitrates are very unlikely to be found as potential ores (since they are soluble in water and would have been washed out of rocks many years ago), they might be made at the first stage in an industrial process so it is worthwhile investigating the ease and extent of their decomposition.

Nitrates of metals decompose in three different ways:

- to oxygen gas and the nitrite
- to oxygen gas, nitrogen dioxide gas and the metal oxide
- to oxygen gas, nitrogen dioxide gas and the metal.

Nitrogen dioxide gas is easily seen since it is dark brown even in small amounts. Oxygen gas is easily detected since it relights a 'glowing splint'.

Figure 4.4
Investigating the action of heat on a nitrate.

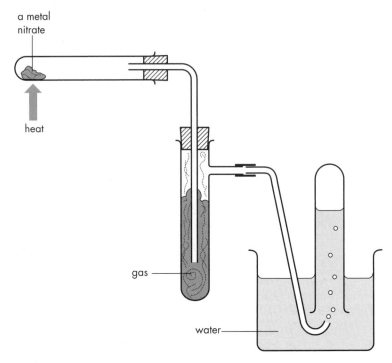

Materials
- eye protection
- test-tubes
- 2 hard glass test-tubes, 125 × 16 mm
- right-angled delivery tube with corks (see Figure 4.4)
- side-arm test-tube
- delivery tube
- small trough
- test-tube holder
- Bunsen burner
- spatula
- wooden splints
- heatproof mat
- samples of potassium nitrate, sodium nitrate, zinc nitrate, lead(II) nitrate, copper(II) nitrate and silver nitrate

Safety
- *Potassium nitrate is oxidising.*
- *Sodium nitrate is oxidising.*
- *Zinc nitrate is oxidising.*
- *Lead(II) nitrate is oxidising, toxic, has a danger of cumulative effects and may harm an unborn child.*
- *Lead(II) nitrate crystals should not be handled by pupils below Year 9.*
- *Copper(II) nitrate is oxidising and is harmful.*

- *Silver nitrate is oxidising, is corrosive and blackens skin.*
- *Nitrogen dioxide gas is very toxic and is corrosive. It is important that pupils are not exposed to nitrogen dioxide. The apparatus used will allow them to see the brown gas, if present, but it dissolves in water, so only oxygen will reach the collecting tube.*
- *Note that hot test-tubes burn skin.*
- *After heating lead(II) nitrate, the test-tube will be weakened.*
- *Keep all dust to a minimum.*
- *Wear eye protection when using a Bunsen burner.*

Procedure
1. Put about 1 cm depth of the nitrate in the test-tube and connect up the apparatus as in Figure 4.4.
2. Heat gently at first then strongly.
3. The first gas collected in the test-tube will be air expanding from the apparatus. Keep heating either until you have three test-tubes of gas or until nothing new seems to be happening. (The solid must be molten in the case of sodium or potassium nitrates.)
4. Arrange the nitrates in order of ease of decomposition and the extent of decomposition.

What you might expect

Table 4.5 *The results of heating some metal nitrates.*

Nitrate	Ease of decomposition	Extent of decomposition	Equation
Potassium nitrate	Difficult	Only oxygen gas produced	$2KNO_3 \rightarrow 2KNO_2 + O_2$
Sodium nitrate	Difficult	Only oxygen gas produced	$2NaNO_3 \rightarrow 2NaNO_2 + O_2$
Zinc nitrate	Needs steady heating	Oxygen gas, brown nitrogen dioxide gas and a yellow solid produced, which changes to white on cooling	$2Zn(NO_3)_2 \rightarrow 2ZnO + 4NO_2 + O_2$
Lead(II) nitrate	Easy	Oxygen gas, brown nitrogen dioxide gas and a yellow solid produced with lots of noise	$2Pb(NO_3)_2 \rightarrow 2PbO + 4NO_2 + O_2$
Copper(II) nitrate	Easy	Oxygen gas, brown nitrogen dioxide gas and a black solid produced	$2Cu(NO_3)_2 \rightarrow 2CuO + 4NO_2 + O_2$
Silver nitrate	Easy	Oxygen, nitrogen dioxide and silver metal produced	$2AgNO_3 \rightarrow 2Ag + 2NO_2 + O_2$

It is probably worth linking this work back to the reactivity series to reinforce the idea that the more reactive the metal is, the harder it is to produce it by decomposing its compounds. The least reactive metals may be simply produced.

Displacing metals with other metals

The practical described earlier (see page 116) could also be used as an economic exercise to show that using an expensive metal to make a cheaper one is uneconomic.

Remind pupils that reduction is the addition of electrons. Hence:

$$Mg + CuSO_4 \rightarrow MgSO_4 + Cu$$

is summarised as

$$Mg + Cu^{2+} \rightarrow Mg^{2+} + Cu$$

or as

$$Mg \rightarrow Mg^{2+} + 2e^-, \quad \text{which is oxidation}$$

and

$$Cu^{2+} + 2e^- \rightarrow Cu, \quad \text{which is reduction.}$$

Reducing an ore with methane

Methane (or natural gas), passed over a hot oxide, is a much more effective reducing agent than many textbooks suggest. There are, however, a number of safety aspects to consider.

The simplest approach is to use a test-tube with a hole blown at the bottom for the escaping methane. Often, the pupils have more fun making the hole in the test-tube than they do performing the experiment.

Materials
- eye protection
- soft glass test-tube, 125 $\times$ 16 mm
- tube from gas tap with cork to fit test-tube
- Bunsen burner and heatproof mat
- boss, stand and clamp
- spatula
- copper(II) oxide, lead(II) oxide, zinc oxide, magnesium oxide or aluminium oxide

Safety
- *Lead(II) oxide is toxic.*
- *Copper(II) oxide is harmful.*
- *Be sure the methane has displaced all the air from the tube before igniting: mixtures of gas and air can be explosive. Collect samples in a test-tube and ignite at arm's length.*
- *Cool the test-tube before touching it.*
- *Keep dust to a minimum.*
- *Wear eye protection when using a Bunsen burner.*

Procedure for making the hole in the test-tube
(This could be done in advance by a technician.)

1. Put a cork or rubber bung in the mouth of a test-tube and hold it there by the pressure of a finger.
2. Heat the bottom of the test-tube with a hot blue Bunsen flame. Once the glass is red-hot, the expanding air will blow a hole in the wall of the test-tube.
3. Allow to cool by leaving on a heatproof mat.

Procedure

1. Using a spatula, put a little copper(II) oxide in the centre of the test-tube and arrange as shown in Figure 4.5, with the test-tube clamped horizontally and with the clamp at the mouth of the test-tube.

Figure 4.5
Reducing copper(II) oxide with natural gas (methane).

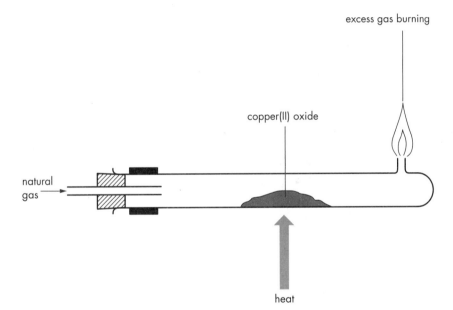

2. Turn on the methane flow but do not light it yet.
3. Allow the methane to flow for about 30 seconds. Light the methane and adjust the gas tap so that the flame is about 2 cm long.
4. Heat the copper(II) oxide and watch what happens. When all reaction seems to have stopped, allow the test-tube to cool, keeping the methane flowing and the flame lit. (This prevents the hot copper from coming into contact with air and re-oxidising.)
5. When cool, tip out the product and examine it to see if it is metallic copper.
6. The experiment may be repeated with lead(II) oxide, zinc oxide, magnesium oxide or aluminium oxide.

What you might expect

When copper(II) oxide is heated in a stream of natural gas a glow should flow through the powder and it should turn from black copper(II) oxide into pink copper metal powder.

A typical equation is difficult to write since the reaction is undoubtedly complicated but it could be represented by

$$4CuO + CH_4 \rightarrow 4Cu + CO_2 + 2H_2O$$

and summarised as

$$Cu^{2+} + 2e^- \rightarrow Cu \quad \text{(i.e. reduction)}$$

Lead(II) oxide will be reduced to lead, but the other oxides will not be reduced.

Reducing an ore with carbon

Carbon blocks are not as common as they once were in schools, and their use for the reduction of oxides, using a blow torch, is hazardous because of the dust produced. You are advised to use the alternative approach described below.

Materials
- eye protection
- crucible
- pipe-clay triangle
- tripod
- Bunsen burner and heatproof mat
- hydrochloric acid (1 mol/dm^3 or 0.5 mol/dm^3)
- carbon powder
- copper(II) oxide, lead(II) oxide, iron(III) oxide, zinc oxide, magnesium oxide

Safety
- *Lead(II) oxide is toxic.*
- *Copper(II) oxide is harmful.*
- *Ventilate the room well.*
- *Wear eye protection when using a Bunsen burner.*

Procedure
1. Mix copper(II) oxide with carbon powder thoroughly. Do not produce a lot of dust.
2. Place about 1 cm depth of the mixture in a crucible and heat strongly.
3. Allow to cool on a heatproof mat.
4. When cool, investigate whether or not any metal has been produced – either by inspection or by reaction with dilute hydrochloric acid.
5. The experiment can be repeated with lead(II) oxide, iron(III) oxide, zinc oxide and magnesium oxide.

<u>What you might expect</u>
Reduction to metals will occur with all except magnesium oxide:

$$2CuO + C \rightarrow 2Cu + CO_2$$
$$2PbO + C \rightarrow 2Pb + CO_2$$
$$2Fe_2O_3 + 3C \rightarrow 4Fe + 3CO_2$$
$$2ZnO + C \rightarrow 2Zn + CO_2$$

Reducing an ore with electricity

By now pupils should have the idea that reduction is the addition of electrons, and they might suggest that electrons could be added using electricity.

The first problem to be overcome, however, is that the starting material (known as the 'electrolyte') must be a liquid. Industrially, it is relatively easy to melt large quantities of a compound and then pass electricity through the melt but this is less practical in a school situation.

Begin with a series of liquids made by dissolving compounds in water.

Producing a metal by electrolysis
<u>Materials</u>
- eye protection
- 100 cm^3 beaker
- 2 carbon rods
- 2 cables with crocodile clips at each end
- 6–8 V power pack
- copper(II) sulphate solution (1 mol/dm^3)
- magnesium sulphate solution (1 mol/dm^3)

<u>Safety</u>
- *Copper(II) sulphate solution is harmful if swallowed and may be irritating to the skin.*
- *Do not touch the power pack with wet hands.*
- *Do not let the carbon electrodes touch each other.*
- *Wear eye protection when carrying out electrolysis.*

<u>Procedure</u>
1. Pour about 2 cm depth of the solution into a beaker.
2. Put in place two carbon rods as electrodes and use the cables to connect them to a power pack (see Figure 4.6).
3. Pass a current through the circuit at a voltage of around 6 to 8 volts.
4. After about 3 minutes, switch off and remove the carbon rods. Inspect the cathode.

Figure 4.6

The electrolysis of copper(II) sulphate solution.

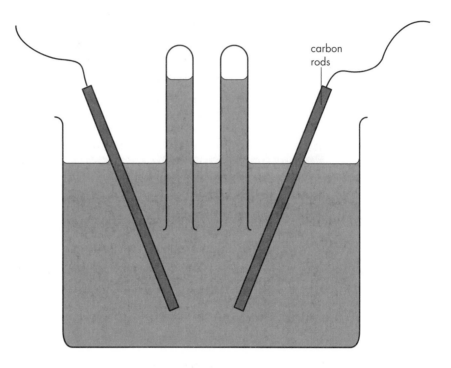

carbon rods

Since pupils are not studying electrolysis at this point, there is no need to investigate in depth all that is happening. The formation of a metal is a reduction process where electrons are added to a positive ion to make a neutral atom:

$$Cu^{2+} + 2e^- \rightarrow Cu$$

This takes place at the cathode – the negative electrode. (An image of electrons being pumped down the cathode and up the anode is sufficient to make the point.)

The complication involved in using a solution of a compound in water is that the water also has positive ions in it. About one in every 10 million water molecules splits up (dissociates) into a pair of ions:

$$H_2O \rightarrow H^+ + OH^-$$

This doesn't seem like much until you appreciate that every 10 cm^3 of water contains about 3×10^{23} molecules. So, if one in 10^7 water molecules splits into ions there are 3×10^{16} H$^+$ ions.

If the solution contains 1 mole of CuSO$_4$ per dm^3, that is 6×10^{23} copper ions per 1000 cm^3, or 6×10^{21} copper ions per 10 cm^3. So the copper ions outnumber the hydrogen ions in the ratio 200 000 : 1.

The work on the reactivity series done so far suggests that copper is easier to produce from its ions than hydrogen is, and the copper ions are more common in this solution, so it is not too surprising that copper is produced at the cathode. Using magnesium sulphate solution does not produce magnesium but gives hydrogen instead, because that is so much 'easier'.

Note that the experiment works for zinc even though the theory says it 'shouldn't', so it is probably less confusing to pupils if you use only copper(II) sulphate and magnesium sulphate solutions. You do not need to introduce redox potentials and over-voltages unless you really wish to extend your pupils, and even then there is a danger that they will lose sight of the overall theme of the unit.

A brief class discussion should lead easily to the conclusion that the process of producing a metal by electrolysis would be much simpler if it were done not in solution but in a liquid electrolyte made by simply melting the compound.

Demonstration of the electrolysis of molten zinc chloride

Materials
- eye protection
- 2 crucibles
- pipe-clay triangle
- tripod
- Bunsen burner
- tongs
- power pack, 12 V
- 2 carbon rods
- 2 cables with crocodile clips at each end, one clip holding a bulb
- spatula
- zinc chloride

Safety
- *Zinc chloride is corrosive.*
- *This experiment must be done as a demonstration by the teacher in a fume cupboard as toxic chlorine vapour is produced.*
- *In many books lead(II) bromide is recommended. However, lead(II) bromide vapour is toxic and zinc chloride is easier to melt than lead(II) bromide.*
- *Wear eye protection when using a Bunsen burner.*

Procedure

Figure 4.7
*Obtaining zinc
from molten zinc
chloride by
electrolysis.*

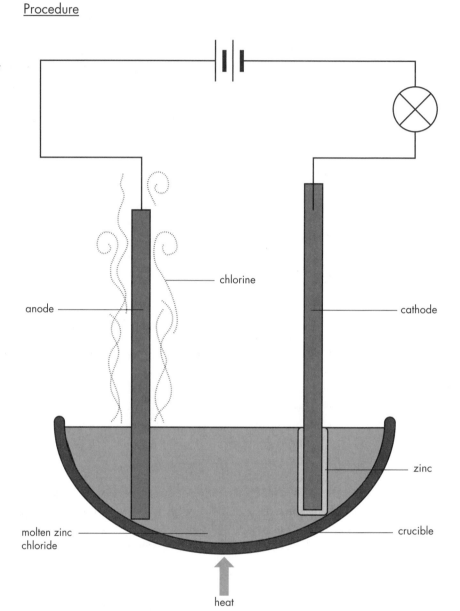

1.   Set up the apparatus as in Figure 4.7. Note that the carbon rods
     should reach to the bottom of the crucible and the crucible should be
     full of solid at the beginning, since the volume of the solid shrinks
     considerably on melting.
2.   Set the voltage to 12 volts and switch on the power pack. Show that
     there is no current flowing through the solid.

3. Heat the crucible until the zinc chloride melts and show that the bulb now lights.
4. Pass the electricity, at about 2 amps, for about 15 minutes. During this time the class could individually inspect the experiment, from a safe distance, and note the greenish-yellow fumes of chlorine gas coming off (toxic). These are best seen against a white background.
5. After 15 minutes switch off the current. Using tongs, carefully pour the contents of the crucible into another crucible, leaving a small pool of molten zinc metal behind in the first one.
6. Allow to cool and then show the class the contents of the first crucible.

What you might expect
Before starting the experiment, check that the circuit is complete by touching both carbon rods at the same time with a metal spatula; the bulb should light. If it does not, check the connections in the circuit and the bulb.

At the start of the experiment the bulb is not lit. As the zinc chloride melts, the bulb lights. Immediately this happens, remove the heat and let the zinc chloride solidify; the bulb should go out. Then continue the experiment.

The pupils should see greenish chlorine gas leaving the crucible. At the end of the experiment a silvery bead of zinc metal should remain.

Sometimes the bulb continues to glow brightly at the end of the experiment when the zinc chloride has solidified. If this happens it is because the zinc bead has short-circuited the two electrodes: metals are good conductors of electricity.

The only ions present in the molten liquid came from the zinc chloride, so there are no complications of competition, as with solutions, and no concerns about the reactivity series since the ions are being forced to accept electrons (whether they like it or not!):

$$Zn^{2+} + 2e^- \rightarrow Zn$$

By now it should be clear to pupils that:

- less reactive metals could be produced by reduction of their compounds by carbon, etc.
- more reactive metals will need to be made by reduction of their compounds by electrolysis
- expensive metals could be made by displacement from their compounds by cheaper metals (so long as these are more reactive).

4.3 Extraction of common metals

◆ *Previous knowledge and experience*

Pupils are likely to have come across ideas about the extraction of metals in History and Geography, for example, iron in a study of the Iron Age.

Pupils will often study the blast furnace for the extraction of iron in Geography and Technology. It is important that liaison with other departments ensures that there is no contradictory teaching.

◆ *A teaching sequence*

Iron

The solid raw materials in the production of iron are iron ore (usually the oxide), coke (a form of impure carbon) and limestone (calcium carbonate). These are added into the top of a very large furnace known as a 'blast furnace' (Figure 4.8).

Figure 4.8
The blast furnace, for extracting iron from iron oxide.

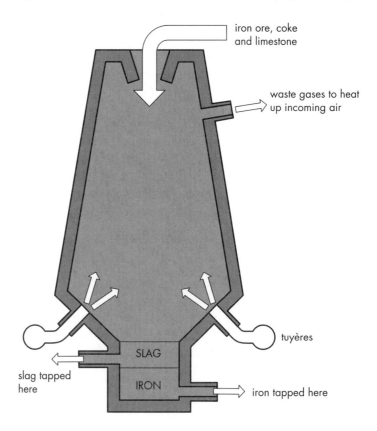

iron ore, coke and limestone

waste gases to heat up incoming air

tuyères

SLAG

IRON

slag tapped here

iron tapped here

Hot air is blown into the bottom of the furnace and this causes the coke to burn, creating a temperature of around 2000 °C. Iron is not produced here.

$$C + O_2 \rightarrow CO_2$$

The carbon dioxide is blown up the furnace where it meets more coke, at a lower temperature. A reaction between the two produces carbon monoxide. Iron is not produced here either.

$$CO_2 + C \rightarrow 2CO$$

The carbon monoxide is blown up the furnace until it meets the iron oxide which has just been added. The carbon monoxide reduces the iron oxide to iron at the top of the furnace.

$$Fe_2O_3 + 3CO \rightarrow 2Fe + 3CO_2$$

As the iron falls down the furnace it becomes hotter, and eventually melts and can be run off from the bottom of the furnace. If any iron oxide falls down the furnace it will be reduced by carbon.

$$Fe_2O_3 + 3C \rightarrow 2Fe + 3CO$$

If *pure* iron oxide were used as the starting material for this process, there would be no need for the limestone. However, purifying the iron oxide would be expensive and the cost would probably not be recovered by increasing the price of the iron produced. So the iron oxide put into the top of the furnace contains various impurities, of which the most significant is silicon dioxide or sand. The limestone which is added is used to remove this silicon dioxide. The limestone falls to half-way down the furnace, where it decomposes.

$$CaCO_3 \rightarrow CaO + CO_2$$

The CO_2 produced is blown up the furnace, reduced to CO and reacts with the iron ore. The hot CaO, calcium oxide, reacts with the sand to form calcium silicate, which almost melts to form a 'liquid' a little like ice-cream (but a lot hotter!).

$$CaO + SiO_2 \rightarrow CaSiO_3$$

This is called 'slag' and is run off from on top of the molten iron at the bottom of the furnace.

The final iron is called 'cast iron' and is very impure. It is purified, mainly by melting it and blowing oxygen through it to burn away all the impurities (as oxides). Then measured amounts of elements such as carbon, chromium or manganese are added, forming mixtures known as 'steels'.

Zinc

Zinc is extracted from zinc oxide by heating a mixture of zinc oxide and carbon, in a similar process. Pupils could research details of its extraction. Because of the low boiling point, the zinc vapour escapes from the furnace and is collected as a liquid by condensation. The zinc collected is quite pure because the impurities are not volatile and remain in the furnace.

Aluminium

The raw material for the production of aluminium is bauxite (aluminium oxide). Aluminium is too reactive a metal to be produced by reduction with carbon at the temperatures typical of the blast furnace. It would take temperatures in excess of 2500 °C to produce aluminium in a blast furnace and this gives three major problems:

1. The furnace would have to be made resistant to these temperatures.
2. The aluminium produced would be a gas – although the zinc production method deals with that problem, it does not have to deal with these temperatures.
3. The aluminium reacts with carbon at these temperatures to form aluminium carbide, so little actual metal would be made.

Note that it is not true to say that aluminium cannot ever be reduced by carbon.

So we are left with electrolysis (see Figure 4.9). Aluminium oxide is a very poor conductor of electricity and has a high melting point, so to make a conducting solution which will melt at about 600 °C cryolite (Na_3AlF_6) is used as a solvent. This concept is a difficult one for pupils to handle, since cryolite is a rock. Obviously, when melted the rock becomes a liquid which can be used as a solvent.

Do not tell pupils that cryolite 'lowers the melting point of the aluminium oxide' since that is impossible – Al_2O_3 has a melting point which is fixed at atmospheric pressures, and doesn't change much at other pressures. Rather, the cryolite 'dissolves the oxide to give a conducting solution which is molten at a lower temperature than would be true for the oxide alone'.

Figure 4.9
*Electrolysis cell
for extracting
aluminium from
bauxite
(aluminium
oxide).*

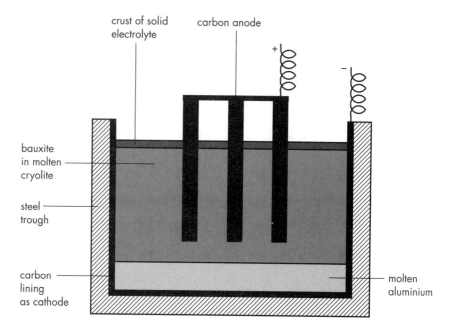

The electrodes are made of carbon (graphite), and the
aluminium forms at the cathode.

$$Al^{3+} + 3e^- \rightarrow Al$$

The oxygen produced at the anode

$$2O^{2-} - 4e^- \rightarrow O_2$$

causes the carbon anode to burn away to carbon dioxide, so
the anodes have to be replaced regularly.

Pupils can find out a great deal about the process of
extracting aluminium from aluminium oxide, and could discuss
why a smelter is sited in a particular place. Then a role play
exercise could be set up. Suppose an aluminium manufacturer
wants to set up a new smelter in your area.

- What would be the views of different people in the
  community?
- What would the firm be looking for in a new site?
- What problems could the smelter cause to the environment?

You could then stage a planning enquiry, with different pupils
taking different roles.

◆ *Enhancement ideas*

- ◆ Pupils could make a model blast furnace out of clay.

4.4 Purification of copper

♦ *A teaching sequence*

Copper is made by a process similar to the blast furnace but, as with iron, it is collected containing lots of impurities. To purify the copper, it is made the anode of a very large electrolysis cell containing copper sulphate solution as the electrolyte and a pure copper rod as the cathode. The anode dissolves, depositing all the impurities on the floor of the cell, and pure copper is deposited onto the cathode. The copper made this way is about 99.999% pure. The process can be illustrated in class by a simple electrolysis experiment.

Purifying copper by electrolysis
Materials
- eye protection
- beaker
- carbon rod
- piece of copper foil
- 2 cables with crocodile clips at both ends
- power pack, 6–8 V
- copper(II) sulphate solution (1 mol/dm^3)

Safety
- *Copper(II) sulphate solution is harmful if swallowed and may be irritating to the skin.*
- *Do not touch the power pack with wet hands.*
- *Wear eye protection when carrying out electrolysis.*

Procedure
1. Pour about 2 cm depth of the copper(II) sulphate solution into the beaker.
2. Put in place a carbon rod as the cathode (negative) and a piece of copper foil as the anode (positive) and connect, via crocodile clips and cables, to a power pack (see Figure 4.10).
3. Pass a current through the circuit at a voltage of around 6 to 8 volts.
4. After about 3 minutes, switch off and remove the carbon rod; it should be coated with copper.
5. If you replace the cathode, reconnect and pass a current for about 15 minutes, eventually the anode will 'dissolve' away.

At the cathode: $Cu^{2+} + 2e^- \rightarrow Cu$ (reduction)
At the anode: $Cu \rightarrow Cu^{2+} + 2e^-$ (oxidation)

Figure 4.10
Purifying copper by electrolysis.

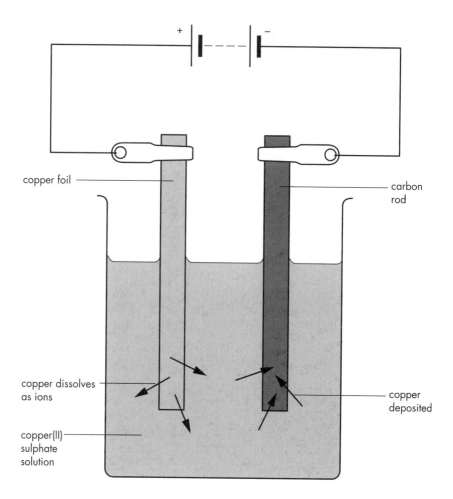

◆ *Enhancement ideas*

- ◆ Metals can be plated by other metals, either by chemical displacement or by using electrolysis. A piece of copper left in silver nitrate solution will be silver plated, as copper is more reactive than silver.

  $Cu + 2AgNO_3 \rightarrow Cu(NO_3)_2 + 2Ag$

- ◆ A piece of brass can be zinc plated using zinc sulphate solution as the electrolyte and the piece of brass as the cathode. Electroplating is used to prevent corrosion and to decorate metals.
- ◆ Anodising aluminium and dyeing the anodised aluminium (see *Other resources*, page 140).

4.5 Ores of metals

♦ *Previous knowledge and experience*

Pupils will have some knowledge of rocks from primary school and will know that rocks are a source of materials.

♦ *A teaching sequence*

Igneous rocks were made millions of years ago as the molten mass from the centre of the Earth escaped to the surface, or nearly, and cooled. Sedimentary rocks were formed as layers of sediments in rivers or seas were deposited on top of each other and squeezed by the pressure. Metamorphic rocks are formed from the others by high pressure and high temperature. In all cases, the rocks are there and there is nothing we can do about it!

Over the years many rocks have weathered and various ions have been lost from their structure; the rock becomes simpler. A typical change, over millions of years, found in hot dry countries would be $NaAlSi_3O_4$ weathering to become $Al_2O_3.3H_2O$.

We have to use the rocks that are available. The key feature is the concentration of the rock and its economic viability. For example, suppose that copper metal can be sold in the UK at £1000 a tonne and that an ore has been found in a central African country which contains 1% copper carbonate, $CuCO_3$. 124 tonnes of copper carbonate gives 64 tonnes of copper, so making 1 tonne of copper requires 1.9 tonnes of copper carbonate or 190 tonnes of this ore. The costs might add up like this:

Mining the rock @ £1 per tonne of rock	£190
Tax @ 10p per tonne of rock	£19
Purifying and extracting the $CuCO_3$ from the rock @ £1 per tonne of rock	£190
Transporting 1.9 tonnes of $CuCO_3$ to the coast and then to the UK @ £30 per tonne	£57
Smelting 1.9 tonnes of $CuCO_3$ to 99% pure copper @ £100 per tonne	£190
Electrolytic refining of 1 tonne of copper to 99.999% pure @ £120 per tonne	£120
Total	£766

 This calculation could be done as a spreadsheet and used to discuss how companies could lose money or make a lot, how the African country could reap more benefit, and so on. For further discussion see *Science in Society Teacher's Guide*, project director: John L. Lewis, 1981, ISBN 0 435 54043 2. Heinemann Educational, London.

Many ores are oxides (e.g. iron oxide and aluminium oxide), some are carbonates (e.g. copper carbonate) and some are sulphides (e.g. lead sulphide and zinc sulphide). Although pupils are not expected to know the names of a large number of ores, Table 4.6 may be useful for reference.

Table 4.6 *Some ores used for the extraction of metals.*

Name of ore	Chemical formula	Metal extracted
Haematite	Fe_2O_3	Iron
Bauxite	Al_2O_3	Aluminium
Halite	$NaCl$	Sodium
Zinc blende (sphalerite)	ZnS	Zinc
Malachite	$Cu_2CO_3(OH)_2$	Copper
Copper pyrites (chalcopyrite)	$CuFeS_2$	Copper
Ilmenite	$FeTiO_3$	Titanium
Wolframite	$FeWO_4$	Tungsten
Cassiterite	SnO	Tin
Galena	PbS	Lead

◆ *Enhancement ideas*

◆ Pupils could look in daily newspapers, e.g. the *Financial Times*, to find the prices of metals.

◆ *Other resources*

The video *Industrial Chemistry for Schools and Colleges* and background information from the Royal Society of Chemistry (see page 297) contains resources on copper refining, iron and steel production and sodium production by electrolysis of molten sodium chloride.

The Aluminium Federation (Broadway House, Calthorpe Road, Five Ways, Birmingham B15 1TN) provided all schools with a series of OHP transparencies on the extraction process.

Information for teachers and pupils on iron and steel can be obtained from British Steel Educational Service, PO Box 10, Wetherby, W Yorks LE23 7EL.

Information on copper and its purification can be obtained from the Copper Development Association, Verulam Industrial Estate, 224 London Road, St Albans, Herts AL1 1AQ.

Rio Tinto is a world leader in finding and processing the Earth's mineral resources. Information is available from Rio Tinto plc, 6 St James's Square, London SW1Y 4LD.

Classic Chemistry Demonstrations (RSC, 1998, ISBN 1 870343 38 7) was circulated free to all UK secondary schools. Demonstrations relevant to this section are:

  5 The combustion of iron wool
 11 The reaction between zinc and copper oxide
 17 Anodising aluminium
 18 The real reactivity of aluminium
 43 Movement of ions during electrolysis
 49 The reaction of sodium with chlorine
 53 Reduction of copper oxide
 74 The thermit reaction
 75 The reaction of magnesium with steam
 84 Zinc plating copper and the formation of brass
    – turning copper into silver and gold
 85 The electrolysis of lead bromide

Displacement reactions of metals can be carried out on a microscale (see *Microscale Chemistry* No. 5, RSC, 1998, ISBN 1 870343 49 2).

Good coverage of this section, and corrosion in particular, is given in *Metals and Corrosion*, ISBN 0 7487 0211 3, by Bob McDuell, published by Stanley Thornes, Old Station Drive, Leckhampton, Cheltenham GL53 0DN.

 Some of the chemical reactions in this section are shown on the CD ROM *The Elements* from Yorkshire International Thomson Media (see Chapter 2).

 ## Web sites

♦ A photograph of the blast furnace for extracting iron can be found on:
  www.heprefs.co.uk/blastfce.html
♦ A website on metal ores is:
  www.greentie.org/class/eoc13.htm
♦ The British Steel home page provides information about iron and steel production in the UK. This can be found on:
  www.britishsteel.co.uk
♦ Aluminium production can be found on:
  www.xisltd.co.uk/alfed/alfhome.html
  This web page gives good environmental information about the process.
♦ Another web page for aluminium production is:
  www.aia.aluminium.qc.ca/english/partenaire/ alcan-a.html

5 *Chemical calculations and the mole*

Judith Johnston

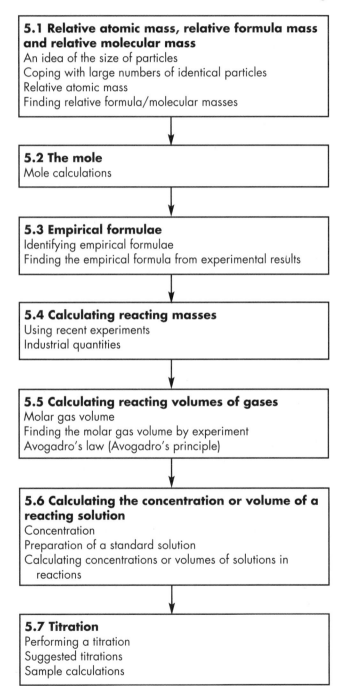

5.1 Relative atomic mass, relative formula mass and relative molecular mass
An idea of the size of particles
Coping with large numbers of identical particles
Relative atomic mass
Finding relative formula/molecular masses

5.2 The mole
Mole calculations

5.3 Empirical formulae
Identifying empirical formulae
Finding the empirical formula from experimental results

5.4 Calculating reacting masses
Using recent experiments
Industrial quantities

5.5 Calculating reacting volumes of gases
Molar gas volume
Finding the molar gas volume by experiment
Avogadro's law (Avogadro's principle)

5.6 Calculating the concentration or volume of a reacting solution
Concentration
Preparation of a standard solution
Calculating concentrations or volumes of solutions in
   reactions

5.7 Titration
Performing a titration
Suggested titrations
Sample calculations

♦ *Choosing a route*

The teaching of the mole concept (with all that has to be established first in a logical way before the concept can be properly understood) has always presented a major challenge to both teachers and pupils. As with all major concepts, it is not possible to teach the mole concept effectively so that pupils understand it and can make use of it, if it is taught as a single, separate topic. Even for able pupils, the growth of understanding takes time.

The original Nuffield Chemistry course in the 1960s, and later revised versions, showed how to 'drip-feed' the teaching of this concept over two or three years, and there is no doubt that this is an effective way to approach it. In these days of modular courses, however, there is a strong tendency to teach the concept as a single intensive topic and, as a consequence, few pupils gain real mastery of it by the end of a GCSE course. Often the topic is left until late in the course so that there is little time for understanding to develop with practice. Even at A level, many pupils still lack confidence with the concept.

If the concept is to be taught effectively, it is essential that the underpinning concepts (see *Previous knowledge and experience* below) have been understood. The 'Children's Learning in Science' project (CLIS) was a government funded follow up to the findings of the Assessment of Performance Unit. CLIS was based at Leeds University Centre for Science Education and researched pupils' problems in understanding a simple particulate view of matter in depth, and demonstrated effective teaching and learning approaches for 12–13 year olds. At the same time, the concepts of elements and compounds need to be considered before pupils are in a position to grasp the concept of an atom. However, the learning of chemical symbols (and even some simple chemical formulae) as a rote exercise can give these younger pupils more confidence when the more conceptual stages are met at age 14 to 15. Similarly, 'word equations' as descriptions of what happens are a useful way of summarising chemical changes up to that age. Some more able pupils, who can assimilate symbolic approaches, may also take on board symbolic chemical equations, and even balance them according to a set of rules, but it is unrealistic to think they have much understanding of what they are doing.

The mathematics involved divides into two stages. There needs to be confidence with the basic arithmetic functions applied to simple numbers (including decimals) before pupils can move on to matters such as standard form and the use of ratios. Liaison with the Mathematics department is essential if Science teachers are not to make unreasonable assumptions about the mathematical skills of most 14–15 year olds.

So to the development of the mole concept over the two years to GCSE using a drip-feed approach. The idea of a chemical formula to represent the composition of a chemical compound is best taught early in this phase. The approach can be experimental, but practice is also needed with calculations that lead from reacting quantities to chemical formulae. The idea of molecules needs to be established at this stage too, especially in respect of the gaseous elements. This is also the time to establish the terminology of relative formula masses.

The move from formulae to simple equations for the reactions of two elements to form a compound provides an easy development. If a few weeks have elapsed since pupils had their first encounter with formula calculations, then they will not feel so overwhelmed. Similarly, further developments to the gas laws and reacting volumes, to ionic equations for precipitation or ionic displacement reactions, and hence to the use of calculations based on concentrations, need to be spaced out over the two years. For less able pupils making slower progress, it is more important to reinforce the concepts they have already tried to master with limited success, than to push ahead into more difficult conceptual areas on an unsound foundation.

You might prefer to teach this topic in separate chunks. Section 5.1 could be taught with the atom (Chapter 2) and the difference between mass number and relative atomic mass can be made clear; section 5.2 might also be suitable here because of the work on the size of the particles. You might consider entering section 5.3 via combustion of magnesium. Section 5.4 could be introduced using any suitable reactions, and section 5.5 might be combined with any work on gases, possibly before or after the gas equation ($PV/T =$ constant). You might care to teach sections 5.6 and 5.7 as one topic.

5.1 Relative atomic mass, relative formula mass and relative molecular mass

♦ *Previous knowledge and experience*

Pupils attempting quantitative work on the mole should know:

- that all matter is made of small particles – atoms, molecules or ions (see Chapter 2)
- chemical symbols and formulae (see Chapter 1).

They should be able to add, subtract, multiply and divide simple numbers. They should understand simple ratios and be able to write large and small numbers in standard form.

♦ *A teaching sequence*

Atoms are too small to be weighed or counted separately. It is very difficult for anyone to appreciate how tiny atoms are and you might find that it can be useful to spend some time on this. This section links with work covered in Chapter 2.

An idea of the size of particles

1. Show the pupils a cardboard box, or other suitable container, with a volume of 24 dm^3 (e.g. a cube of side 29 cm). Tell the pupils that it holds 6×10^{23} particles of air gases. Ask pupils to write out this number as 6 followed by 23 noughts and see if they can put it into words. It is important that they appreciate just how large this number is, and a couple of examples may help:

   - If the whole population of the world wished to count up to this number between them and they all worked at counting without any breaks at all, it would take six million years for them to finish.
   - A line 6×10^{23} mm long would stretch from the Earth to the Sun and back two million times.

   If the box contains so many particles, they must be very small!

2. Have weighed out and on display 12 g of charcoal, 18 g of water, 64 g of copper, 56 g of iron and as many other examples of a mole as resources allow. Each of these samples contains an equal number of particles, namely 6×10^{23}.

Coping with large numbers of identical particles

Particles are too small to weigh or measure individually.

1. Provide a large number of coins of the same value (and size). Ask the pupils to count them *quickly*. Ask for suggestions for another method. With prompting, if necessary, get to the idea of finding the mass of one coin, weighing all the coins and then calculating the number of coins.
2. Repeat this, using dried peas, beads, buttons or playing cards.

Relative atomic mass

The relative atomic mass is the mass of an atom compared with (relative to) one-twelfth of the mass of an atom of carbon-12. It is a weighted average of all the naturally occurring isotopes of the element.

The idea of using a relative mass instead of a real one for very small objects is not necessarily easy for the pupils to grasp. You may care to spend some time developing their understanding by using exercises such as those which follow.

Using relative masses

1. Use lumps of prepared Plasticine in different sizes, for example:

smallest lumps	0.82 g
small lumps	1.64 g
medium lumps	2.46 g
large lumps	3.28 g

    The pupils have to compare their masses. Ask the pupils if values of 1, 2, 3 and 4 would be easier to handle. This will result in having a '1 unit' lump, a '2 unit' lump, etc.
2. You could also use a collection of buttons of many different sizes with the higher ability pupils.

Finding relative formula/molecular masses

The relative formula mass (RFM) or the relative molecular mass (RMM) is the sum of the relative atomic masses of all the atoms in one formula unit or molecule of the substance; it has no units. To be pragmatic, the use of relative *formula* mass or relative *molecular* mass depends on what is required by your examination board. It seems a pity, however, to spend time emphasising that only covalent substances exist as molecules and then ask a pupil to calculate the relative *molecular* mass of an *ionic* compound!

The calculations can be done by following the 'shape' of the formula. You may prefer, or your pupils may prefer, to reorganise the information in a formula before calculating the RFM or RMM.

Example: $CH_3CO_2C_2H_5$

1. Following the formula 'shape':

   $$12 + (1 \times 3) + 12 + (16 \times 2) + (12 \times 2) + (1 \times 5) = 88$$

2. Rearranging the formula as $C_4H_8O_2$:

   $$(12 \times 4) + (1 \times 8) + (16 \times 2) = 88$$

Sample calculations

You may find that it is useful to start with very simple examples, where there is only one atom of each element, and progress to those with more than one atom of some elements, then to brackets and finally to hydrated compounds.

1. Potassium hydroxide, KOH (RAM: K = 39, O = 16, H = 1) has the RFM

   $$39 + 16 + 1 = 56$$

2. Carbon dioxide, CO_2 (RAM: C = 12, O = 16) has the RFM

   $$12 + (16 \times 2) = 44$$

3. Calcium hydroxide, $Ca(OH)_2$ (RAM: Ca = 40, O = 16, H = 1) has the RFM

   $$40 + 2(16 + 1) = 74$$

   Rearranging the formula as CaO_2H_2, the RFM is

   $$40 + (16 \times 2) + (1 \times 2) = 74$$

4. Ammonium sulphate, $(NH_4)_2SO_4$ (RAM: N = 14, H = 1, S = 32, O = 16) has the RFM

   $$2[14 + (1 \times 4)] + 32 + (16 \times 4) = 132$$

   Rearranging the formula as $N_2H_8SO_4$, the RFM is

   $$(14 \times 2) + (1 \times 8) + 32 + (16 \times 4) = 132$$

5. Cobalt chloride-2-water, $CoCl_2.2H_2O$ (RAM: Co = 59, Cl = 35.5, H = 1, O = 16) has the RFM

   $$59 + (35.5 \times 2) + 2[(1 \times 2) + 16] = 166$$

   Rearranging the formula as $CoCl_2H_4O_2$, the RFM is

   $$59 + (35.5 \times 2) + (1 \times 4) + (16 \times 2) = 166$$

5.2 The mole

◆ *Previous knowledge and experience*

Pupils will have no previous knowledge of the mole from lower-school Science. Appreciation of very large numbers will be useful in understanding this topic. They should be able to add, subtract, multiply and divide simple numbers. They should understand simple ratios and be able to write large and small numbers in standard form.

◆ *A teaching sequence*

The mole is an amount of a chemical. The mole contains 6×10^{23} particles (atoms, molecules or ions as appropriate). 6×10^{23} is the Avogadro constant (also known as Avogadro's number).

Mole calculations

1. How many particles are there in 2 moles of carbon?

   Number of particles = moles × Avogadro constant
   $$= 2 \times 6 \times 10^{23}$$
   $$= 1.2 \times 10^{24}$$

2. How many particles are there in 0.25 moles of oxygen?

   Number of particles = moles × Avogadro constant
   $$= 0.25 \times 6 \times 10^{23}$$
   $$= 1.5 \times 10^{23}$$

3. How many moles of nitrogen contain 3×10^{23} particles?

   $$\text{Number of moles} = \frac{\text{particles}}{\text{Avogadro constant}} = \frac{3 \times 10^{23}}{6 \times 10^{23}}$$
   $$= 0.5$$

4. How many moles of chlorine contain 1×10^{23} particles?

   $$\text{Number of moles} = \frac{\text{particles}}{\text{Avogadro constant}} = \frac{1 \times 10^{23}}{6 \times 10^{23}}$$
   $$= \tfrac{1}{6} \text{ or } 0.17$$

One mole of an element that exists as atoms, weighs the same as its relative atomic mass (RAM) in grams. (This amount used to be called one 'gram-atom' of the element.)

One mole of a compound (or an element that exists as molecules) weighs the same as its relative formula mass (RFM) or relative molecular mass (RMM) in grams. (This amount used to be called one 'gram-molecule' of the compound.)

The molar mass of an element or compound is defined as the mass, in grams, of one mole of an element or compound. Its units are g/mol. The term 'molar mass' is sometimes confused with relative formula or molecular mass. Depending on your syllabus you may prefer not to mention it except to the most able pupils.

The word 'molar' has been used in the past as a unit of concentration, e.g. 2 molar (2 M) hydrochloric acid solution. Now it is used as an adjective, meaning 'of the mole'.

Building up to 'moles = mass/RFM'

Rather than just giving your pupils the formula 'moles = mass/RFM', you might care to encourage them to work it out for themselves.

1 mole of carbon dioxide	(RFM 44)	weighs 44 g
2 moles of carbon dioxide	(RFM 44)	weigh . . .
0.5 moles of carbon dioxide	(RFM 44)	weighs . . .
. . . moles of carbon dioxide	(RFM 44)	weighs 4.4 g
1 mole of water	(RFM 18)	weighs . . .
. . . moles of water	(RFM 18)	weighs 4.5 g
2 moles of X	(RFM . . .)	weigh 73 g
0.5 moles of Y	(RFM . . .)	weighs 40 g
etc.		

Having established moles = mass/RFM, pupils can then be asked to express first mass and then RFM in terms of the other two variables. They should come up with the expressions:

$$\text{Moles} = \frac{\text{mass}}{\text{RFM}}$$

$$\text{Mass} = \text{moles} \times \text{RFM}$$

$$\text{RFM} = \frac{\text{mass}}{\text{moles}}$$

If your pupils are used to working with 'magic triangles', you could introduce one at this point.

Sample calculations

1. How many moles are in 4.0 g of sodium hydroxide, NaOH?

$$\text{Moles of NaOH} = \frac{\text{mass}}{\text{RFM}} = \frac{4.0}{40} = 0.1$$

2. What is the mass of 0.02 moles of calcium carbonate, $CaCO_3$?

$$\text{Mass of } CaCO_3 = \text{moles} \times \text{RFM} = 0.02 \times 100 = 2\text{ g}$$

3. What is the RFM of a substance if 0.25 mole weighs 11 g?

$$\text{RFM} = \frac{\text{mass}}{\text{moles}} = \frac{11}{0.25} = 44$$

Moles and the numbers of molecules

With more able pupils you might consider putting this together with earlier work on Avogadro's number.

Sample calculations

1. How many molecules are in 32 g of sulphur dioxide, SO_2?

$$\text{Moles of } SO_2 = \frac{\text{mass}}{\text{RFM}} = \frac{32}{64} = 0.5$$

$$\text{Number of molecules} = \text{moles} \times \text{Avogadro constant}$$
$$= 0.5 \times 6 \times 10^{23}$$
$$= 3 \times 10^{23}$$

2. What is the mass of 2×10^{23} molecules of ozone, O_3?

$$\text{Number of moles} = \frac{\text{particles}}{\text{Avogadro constant}} = \frac{2 \times 10^{23}}{6 \times 10^{23}}$$
$$= \tfrac{1}{3} \text{ or } 0.33$$

$$\text{Mass of } O_3 = \text{moles} \times \text{RFM} = \tfrac{1}{3} \times 48 = 16\text{ g}$$

5.3 Empirical formulae

◆ *Previous knowledge and experience*

Pupils will have no previous knowledge of empirical formulae. They should be able to add, subtract, multiply and divide simple numbers. They should understand simple ratios and be able to write large and small numbers in standard form.

◆ *A teaching sequence*

These are the simplest formulae. They show the elements present and the ratio of their atoms, but not always the real *number* of atoms. For example, CH_2 is an empirical formula showing that the compound contains carbon and hydrogen and that there is one carbon to every two hydrogens, but it could apply to many different compounds, e.g. C_2H_4, C_3H_6 and $C_{21}H_{42}$.

Identifying empirical formulae

At this stage you might like to give your pupils several formulae and ask them to pick out the empirical ones (see example 1), and also to give them formulae of compounds and ask them to write the empirical formula of each one (see example 2).

Examples

1. (Empirical formulae in bold.)
    HCl H_2O_2 **C_3H_8** C_4H_8 **C_5H_8** **C_4H_8O**

2. (Empirical formulae in brackets.)
    C_3H_6 (CH_2) $C_{22}H_{46}$ ($C_{11}H_{23}$)
    $C_4H_6O_2$ (C_2H_3O) $C_6H_{12}O_6$ (CH_2O)

Finding the empirical formula from experimental results

Method 1. Using the mole

One approach you may care to use is to start by showing a set of results like those for water in Table 5.1. Thus demonstrate that the ratio of moles is the same as the ratio of the atoms. This could be repeated with as many examples as your inspiration can produce and your pupils' needs require.

Table 5.1 *Numbers of moles of hydrogen and oxygen in samples of water.*

Mass of hydrogen (g)	Number of moles of hydrogen	Mass of oxygen (g)	Number of moles of oxygen
0.2	0.2	1.6	0.1
0.4	0.4	3.2	0.2
0.6	0.6	4.8	0.3
0.8	0.8	6.4	0.4
1.0	1.0	8.0	0.5

The simplest formula of water is H_2O. This means that there are twice as many hydrogen atoms as oxygen atoms in the sample, and twice as many hydrogen atoms as oxygen atoms in each molecule of water.

The desired outcome is a recognition that the ratio of the moles is the same as the ratio of atoms.

Method 2. Without using the mole

Table 5.2 shows the amounts of carbon and oxygen present in some samples of carbon dioxide.

Table 5.2 *Numbers of moles of carbon and oxygen in samples of carbon dioxide.*

Mass of carbon (g)	Mass ÷ RAM of carbon (= number of RAMs)	Mass of oxygen (g)	Mass ÷ RAM of oxygen (= number of RAMs)
1.2	0.1	3.2	0.2
2.4	0.2	6.4	0.4
3.6	0.3	9.6	0.6
4.8	0.4	12.8	0.8

The simplest formula of carbon dioxide is CO_2.

The desired outcome is a recognition that the ratio of the 'number of RAMs' is the same as the ratio of the atoms.

Finding the formula of magnesium oxide by experiment

<u>Materials</u>

- eye protection
- crucible and lid (china crucibles are often of poor quality; metal crucibles may be better)
- pipe-clay triangle
- tripod and heatproof mat
- Bunsen burner
- tongs
- spatula
- access to suitable balance(s) (e.g. top pan, direct-reading balance)
- magnesium ribbon

<u>Safety</u>
- *Magnesium is highly flammable. Teachers must supervise carefully.*
- *Wear eye protection when using a Bunsen burner.*

<u>Procedure</u>
1. Weigh the crucible and its lid.
2. Take about 10 cm of magnesium ribbon. Scrape it with the edge of the spatula to remove any oxide, then fold or coil it up to fit into the crucible.
3. Weigh the crucible, lid and magnesium.
4. Place the crucible containing the magnesium on the pipe-clay triangle. Place the lid on carefully (see Figure 5.1).

Figure 5.1
Finding the formula of magnesium oxide by experiment.

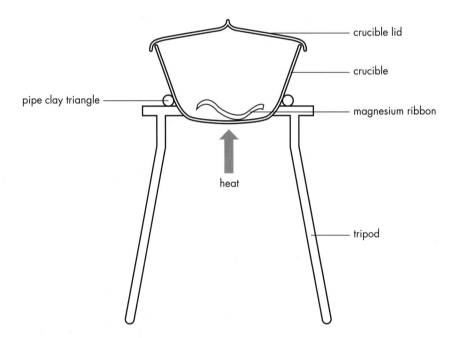

5. Heat the crucible and its contents strongly. Using tongs, tilt the lid a little so that the magnesium can be seen. If it flares up, replace the lid and continue heating. If no flaring is seen, replace the lid and stop heating.
6. Leave to cool.
7. Re-weigh crucible, lid and contents (which should now be a white powder or ash).
8. Repeat steps 5 to 7 until there is no further change in mass.
9. Calculate the mass of magnesium.
10. Calculate the mass of magnesium oxide formed.
11. Calculate the mass of oxygen in the oxide.
12. Calculate the empirical formula of magnesium oxide.

<u>What you might expect</u>
Sample results might be:

mass of crucible and lid	= 17.55 g
mass of crucible, lid and magnesium	= 17.70 g
mass of crucible, lid and magnesium oxide	= 17.80 g
mass of magnesium (17.70 − 17.55)	= 0.15 g
mass of oxygen (17.80 − 17.70)	= 0.10 g

Using the mole, the calculation would proceed thus:

	Mg	**O**
mass	0.15	0.10
RAM	24	16
number of moles	0.00625	0.00625
ratio	1	1
number of atoms	1	1
formula	MgO	

Without using the mole:

	Mg	**O**
mass	0.15	0.10
RAM	24	16
number of RAMs	0.00625	0.00625
ratio	1	1
number of atoms	1	1
formula	MgO	

<u>Possible problems</u>
1. The crucible does not sit steadily on the pipe-clay triangle. The solution is to make sure that they match *before* starting heating.
2. There is no appreciable change in mass. 10 cm of magnesium ribbon weighs approximately 0.15 g and the increase should be about 0.10 g. If this is considered to be too small it is easy to use a longer length, but it will take longer to oxidise it.
3. Some magnesium remains unreacted, so the increase in mass is less than it should be. Unless there is very obviously some metal left it will be hard to tell quickly if the reaction is complete. One possibility is to provide the pupils with ready prepared pieces of magnesium ribbon. You would then know what the increase in mass should be and would be in a position to advise further heating if necessary. (For example, for 0.60 g of magnesium there should be an increase of 0.4 g.)
4. The lid is lifted too high and some magnesium oxide floats out. If this happens, there is no solution other than to start again.
5. The increase in mass is definitely less than expected. This could be due to inefficient removal of any oxide coating on the original sample.

You might like several groups to combine their results and take an average. If there is more time, the groups could each carry out the experiment more than once and take their own average.

This experiment can be repeated using copper powder in the crucible, to form copper oxide.

Heating copper oxide in a stream of hydrogen

You may wish to demonstrate this (see Figure 4.5 on page 125) or to use it as a theoretical example, getting your pupils to work out the formula from a given set of results, for example:

Mass of tube	= 17.24 g
Mass of tube + copper oxide	= 18.04 g
Mass of tube + copper	= 17.88 g

Finding the number of moles of water of crystallisation in a sample of hydrated copper(II) sulphate

The 'water of crystallisation' is the molecules of water that are part of the crystal structure. The number of moles of water of crystallisation present in a hydrated compound can be calculated by a method very similar to that used to find empirical formulae. There are two possible methods of setting out the calculation of the number of moles of water of crystallisation but the experiment is the same for both.

$$CuSO_4.xH_2O \rightarrow CuSO_4 + xH_2O$$

Materials
- eye protection
- dry test-tube, 150 × 16 mm
- Bunsen burner and heatproof mat
- test-tube holder
- access to an accurate balance
- copper(II) sulphate crystals

Safety
- *Copper(II) sulphate crystals are harmful.*
- *Wear eye protection when using a Bunsen burner.*

Procedure
1. Weigh a dry test-tube.
2. Weigh around 2.5 g of copper(II) sulphate crystals accurately into the test-tube.
3. Heat the crystals strongly until there is no further visible change.
4. Cool and re-weigh.
5. Heat the test-tube and contents strongly again for 1 minute.

6. Cool and re-weigh.
7. Repeat steps 5 and 6 until there is no further change in mass.
8. Calculate the mass of anhydrous copper(II) sulphate, i.e. the residue in the test-tube, and hence the mass of water lost from the crystals. Use these figures to calculate the number of moles of water of crystallisation.

<u>What you might expect</u>
Sample results might be:

mass of test-tube	= 21.03 g
mass of test-tube + hydrated copper(II) sulphate	= 23.49 g
mass of test-tube + anhydrous copper (II) sulphate	= 22.61 g
mass of residue (22.61 − 21.03)	= 1.58 g
mass of water of crystallisation (23.49 − 22.61)	= 0.88 g

Using the mole, the calculation might proceed thus:

$$CuSO_4.xH_2O \rightarrow CuSO_4 + xH_2O$$

From the equation:	1 mole	1 mole	x mole
Actual moles from experiment:		$\dfrac{1.58}{160}$	$\dfrac{0.88}{18}$
		= 0.0099	= 0.049
Ratio of moles/molecules		1	4.9

Therefore $x = 4.9$. But there must be a whole number of water molecules so $x = 5$.

The second method is very similar to the set-up for finding an empirical formula:

	$CuSO_4$	**H_2O**
mass	1.58	0.88
RFM	160	18
number of moles	0.0099	0.049
ratio	1	4.9
number of molecules	1	5
formula	$CuSO_4.5H_2O$	

Therefore there are five molecules of water of crystallisation.

Both of these calculation methods can be adapted for use without the mole.

5.4 Calculating reacting masses

◆ *Previous knowledge and experience*

Before attempting this section, pupils should be able to write, or interpret, balanced symbol equations (see Chapter 1).

They should be able to add, subtract, multiply and divide simple numbers. They should understand simple ratios.

◆ *A teaching sequence*

Many books suggest using industry as a starting point. They pose the question, 'How does a manufacturer know how much raw material is needed?' This is fine if your pupils have some notion of industrial processes. You could start by asking if your pupils have any idea how you know how much of any chemical they should use in an experiment.

Using recent experiments

You could take a reaction that you have been looking at recently and write the balanced symbol equation on the board:

$$CuCO_3 \rightarrow CuO + CO_2$$

Then you could ask, 'How did I know that 5 g of copper carbonate would be enough to give a definite decrease in mass?' The reacting masses can be calculated without using the mole by using the RFM.

Sample calculations

1. What mass of sodium chloride will be formed when 4.0 g of sodium hydroxide react with excess hydrochloric acid?

With the mole:

$$NaOH \ + \ HCl \ \rightarrow \ NaCl \ + \ H_2O$$
$$\text{1 mole} \quad \text{1 mole} \quad \text{1 mole} \quad \text{1 mole}$$

$$\text{moles of NaOH} = \frac{\text{mass}}{\text{RFM}} = \frac{4.0}{40} = 0.1$$

From the equation,

$$\text{moles of NaCl} = \text{moles of NaOH} = 0.1$$

Hence,

$$\text{mass of NaCl} = \text{moles} \times \text{RFM} = 0.1 \times 58.5 = 5.85 \text{ g}$$

Without the mole:

$$NaOH + HCl \rightarrow NaCl + H_2O$$

RFMs: 40   58.5

So an actual mass of 40 g would give 58.5 g and the reacted

mass, 4.0 g, would give $\dfrac{58.5 \times 4.0}{40} = 5.85$ g.

2. What mass of silver chloride will be formed when 5.2 g of barium chloride reacts with excess silver nitrate, both in solution?

With the mole:

$$BaCl_2 + 2AgNO_3 \rightarrow 2AgCl + Ba(NO_3)_2$$

1 mole   2 moles   2 moles   1 mole

moles of $BaCl_2 = \dfrac{mass}{RFM} = \dfrac{5.2}{208} = 0.025$

From the equation,

moles of AgCl = moles of $BaCl_2 \times 2$

= 0.025×2

= 0.05

Hence,

mass of AgCl = moles × RFM = 0.05 × 143.5

= 7.2 g

Without the mole:

$$BaCl_2 + 2AgNO_3 \rightarrow 2AgCl + Ba(NO_3)_2$$

RFMs: 208   2 × 143.5

so an actual mass of 208 g would give 287 g and the reacted mass,

5.2 g, would give $\dfrac{287 \times 5.2}{208} = 7.2$ g.

Questions can be asked in which the mass of product required is given and the task is to find the mass of reactant needed.

3. What mass of calcium carbonate, on heating, will give 1.12 g of calcium oxide?

With the mole:

$$CaCO_3 \rightarrow CaO + CO_2$$

1 mole   1 mole   1 mole

moles of CaO = $\dfrac{mass}{RFM} = \dfrac{1.12}{56} = 0.02$

From the equation,

moles of $CaCO_3 = $ moles of CaO = 0.02

Hence,

mass of $CaCO_3 = $ moles × RFM = 0.02 × 100 = 2.0 g

Without the mole:

$$CaCO_3 \rightarrow CaO + CO_2$$

RFMs: 100   56

So an actual mass of 100 g would give 56 g and the given mass,

$$\frac{1.12 \times 100}{56} = 2.0 \text{ g, would give } 1.12 \text{ g.}$$

Industrial quantities

Some questions deal in industrial amounts (e.g. tonnes) and not in grams. The method based on formula masses needs only a change of units. The mole method, however, needs adjustment: ideally, the mass should be changed to grams, but you may prefer to use 'moles' instead, to indicate that these are not 'proper' moles. The advantage of using 'moles' is that the calculated mass will be in the same units as the given mass without any need for conversion.

4. What mass of iron can be extracted from 32 tonnes of iron ore?

With the mole:

$$Fe_2O_3 + \quad 3CO \quad \rightarrow \quad 2Fe \quad + \quad 3CO_2$$

1 mole   3 moles   2 moles   3 moles

$$\text{'moles' of } Fe_2O_3 = \frac{\text{mass}}{\text{RFM}} = \frac{32}{160} = 0.2$$

From the equation,

'moles' of Fe $=$ 'moles' of $Fe_2O_3 \times 2 = 0.2 \times 2 = 0.4$

Hence,

mass of Fe $=$ 'moles' $\times$ RFM $= 0.4 \times 56 = 22.4$ tonnes

Without the mole:

$$Fe_2O_3 + 3CO \rightarrow 2Fe + 3CO_2$$

RFMs: 160   56

So an actual mass of 160 tonnes would give $2 \times 56 = 112$ tonnes and the given mass, 32 tonnes, would give $\dfrac{112 \times 32}{160} = 22.4$ tonnes.

5.5 Calculating reacting volumes of gases

♦ *Previous knowledge and experience*

Before attempting this section of work, students should be able to write (or at least interpret) balanced symbol equations (see Chapter 1).

They should be able to add, subtract, multiply and divide simple numbers. They should understand simple ratios.

♦ *A teaching sequence*

Molar gas volume

One mole of any gas, at room temperature and atmospheric pressure (r.t.p.), occupies $24 \, dm^3$ or $24\,000 \, cm^3$. Room temperature is taken as $20 \, °C$ and atmospheric pressure as $1.01 \times 10^5 \, Pa$. Pascals are the SI unit of pressure:

$$1.01 \times 10^5 \, Pa \quad = \quad 101 \, kPa \text{ (kilopascals)} \quad = \quad 1.01 \times 10^5 \, N/m^2$$
$$= \quad 1 \, atm \text{ (atmosphere)} \quad = \quad 760 \, mm \text{ mercury}$$

Some textbooks refer to the molar gas volume as $22.4 \, dm^3$. This is because it used to be measured at *standard* room temperature and pressure (s.t.p.), i.e. $0 \, °C$ and $1.01 \times 10^5 \, Pa$. Other books may refer to *normal* temperature and pressure; this is the same as standard temperature and pressure.

For gases only:

$$\text{number of moles} \quad = \quad \frac{\text{volume } (dm^3)}{24}$$

$$\text{and volume } (dm^3) \quad = \quad 24 \times \text{number of moles}$$

For calculations without the mole, the RFM in grams of a gas occupies $24 \, dm^3$ at r.t.p.

Sample calculations

1. How many moles of gas A occupy $4.8 \, dm^3$ at r.t.p.?

$$\text{Number of moles} \quad = \quad \frac{\text{volume}}{24} \quad = \quad \frac{4.8}{24} \quad = \quad 0.2$$

2. What volume will 0.01 mole of gas B occupy at r.t.p.?

$$\text{Volume} = 24 \times \text{moles} = 24 \times 0.01 = 0.24 \, dm^3$$
$$= 240 \, cm^3$$

3. What is the maximum volume of carbon dioxide that can be given off on heating 6.2 g of copper carbonate until there is no further change in mass?

$$CuCO_3 \rightarrow CuO + CO_2$$

1 mole   1 mole

$$\text{moles of } CuCO_3 = \frac{\text{mass}}{\text{RFM}} = \frac{6.2}{124} = 0.05$$

From the equation,

$$\text{moles of } CO_2 = \text{moles of } CuCO_3$$
$$= 0.05$$

Hence,

$$\text{volume of } CO_2 = 24 \times \text{moles}$$
$$= 24 \times 0.05$$
$$= 1.2 \, dm^3 \text{ or } 1200 \, cm^3$$

4. What mass of water will be formed when 1.2 dm^3 of oxygen react completely with excess hydrogen?

$$2H_2 + O_2 \rightarrow 2H_2O$$

$$\text{moles of } O_2 = \frac{\text{volume}}{24} = \frac{1.2}{24} = 0.05$$

From the equation,

$$\text{moles of } H_2O = \text{moles of } O_2 \times 2 = 0.05 \times 2 = 0.1$$

Hence,

$$\text{mass of } H_2O = \text{moles} \times \text{RFM} = 0.1 \times 18 = 1.8 \, g$$

Finding the molar gas volume by experiment

Materials
- eye protection
- small conical flask or boiling tube
- gas syringe
- delivery tube and bung
- stand, clamp and boss
- measuring cylinder
- magnesium ribbon
- hydrochloric acid (0.5 mol/dm^3)

Safety
- *Magnesium is highly flammable. Teachers must supervise carefully.*
- *Wear eye protection when using hydrochloric acid.*

Procedure
1. Set up the apparatus as shown in Figure 5.2.

Figure 5.2
Finding the molar volume of gas by experiment.

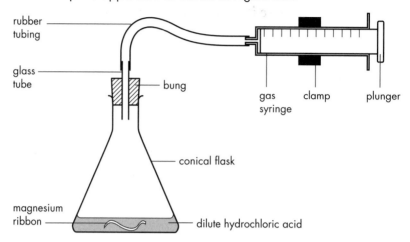

rubber tubing

glass tube

bung

gas syringe   clamp   plunger

conical flask

magnesium ribbon — dilute hydrochloric acid

2. Add 20 cm^3 of hydrochloric acid to the flask.
3. Take a length of the magnesium ribbon and scrape it to remove any oxide coating.
4. Cut exactly 10 cm of the ribbon and weigh it. Use this to calculate the mass of 1 cm by dividing by 10.
5. Cut exactly 1 cm of the ribbon, either from the piece already cut or from a different piece. Coil it so that it will be completely covered by the acid when it is added to the flask.
6. Drop the magnesium into the acid and immediately place the bung in position.
7. When all the magnesium has reacted, note the volume of hydrogen produced.
8. Repeat steps 5 to 7 with different lengths of ribbon (i.e. 2 cm, 3 cm, etc.) up to 5 cm. It is not necessary to add any more acid.
9. Calculate the mass and then the moles of each length of ribbon.
10. Plot a graph of number of moles against volume of hydrogen produced.
11. From the graph find the volume of hydrogen produced by 0.001 mole.
12. Use $Mg + 2HCl \rightarrow MgCl_2 + H_2$ to find the volume of one mole of hydrogen.

What you might expect
Sample results might be:

mass of 10 cm of magnesium ribbon = 0.16 g
From the graph: 0.001 mole of Mg gives 23 cm^3 of H_2.
so   1 mole of Mg gives 23 000 cm^3 (= 23 dm^3) of H_2.
From the equation: 1 mole of Mg gives 1 mole of H_2.
Conclusion: the molar gas volume of hydrogen is 23 dm^3.
                            [Actual value is 24 dm^3.]

Possible problems
The plunger of the gas syringe may not move smoothly; this may occur if the syringe is clamped too tightly.

You may wish to make this a class experiment by giving each group a different length of magnesium and getting them to find the volume more than once and then take an average.

Avogadro's law (Avogadro's principle)

The principle states that 'equal volumes of all gases (at the same temperature and pressure) contain the same number of molecules'. This, combined with the fact that equal numbers of moles contain equal numbers of molecules, means that there is a direct relation between moles and volume for gases:

H_2	+	Cl_2	$\rightarrow$	$2HCl$
1 molecule		1 molecule		2 molecules
1 mole		1 mole		2 moles
1 volume		1 volume		2 volumes

$2CO$	+	O_2	$\rightarrow$	$2CO_2$
2 molecules		1 molecule		2 molecules
2 moles		1 mole		2 moles
2 volumes		1 volume		2 volumes

The ratio of the moles is the same as the ratio of the volumes.

Sample calculations

1. What volume of hydrogen chloride will be formed from 20 cm³ of chlorine reacting completely with hydrogen?

H_2 + Cl_2 $\rightarrow$ $2HCl$
1 mole   2 moles
1 volume   2 volumes
20 cm³   40 cm³

40 cm³ of HCl will be formed.

2. What volume of carbon monoxide is needed to produce 400 dm³ of carbon dioxide by reacting with excess oxygen?

$2CO$ + O_2 $\rightarrow$ $2CO_2$
2 moles   2 moles
2 volumes   2 volumes
400 dm³   400 dm³

400 dm³ of CO is needed.

5.6 Calculating the concentration or volume of a reacting solution

♦ *A teaching sequence*

Concentration

Concentration is the number of moles of solute in 1 dm³ of solution. The units are mol/dm³. It is sometimes given as the mass of solute in 1 dm³ of solution. The units for this are g/dm³. Converting from one concentration unit to another simply requires converting mass to moles or vice versa.

The hazard of a solution depends on concentration, e.g. sulphuric acid solutions less than 0.5 mol/dm³ – low hazard; 0.5–1.5 mol/dm³ – irritant; greater than 1.5 mol/dm³ – corrosive.

Examples

1. A sodium hydroxide solution has a concentration of 0.2 mol/dm³. What is this in g/dm³?

    In 1 dm³ there are 0.2 moles of NaOH.

    So the mass of NaOH in 1 dm³ = moles × RFM
    $$= 0.2 × 40$$
    $$= 8 \text{ g}$$

    Therefore the concentration of the sodium hydroxide solution is 8 g/dm³.

2. A silver nitrate solution has a concentration of 4.25 g/dm³. What is this in mol/dm³?

    In 1 dm³ there are 4.25 g of AgNO₃.

    So the number of moles of AgNO₃

    $$\text{in 1 dm}^3 = \frac{\text{mass}}{\text{RFM}}$$

    $$= \frac{4.25}{170}$$

    $$= 0.025 \text{ mol/dm}^3$$

    Therefore the concentration of silver nitrate is 0.025 mol/dm³.

The volume unit used may be the cubic decimetre (dm^3) or the litre (l). Fortunately, $1\ dm^3 = 1$ litre. It is also important that your pupils can convert from and to the appropriate smaller units, i.e.

$1000\ cm^3 = 1\ dm^3$ and $1000\ ml = 1$ litre

You may wish to give the expression for concentration directly to your pupils or you might prefer to encourage them to work it out for themselves. For example:

1 mole of solute in $1\ dm^3$ has concentration $1\ mol/dm^3$.
2 moles of solute in $1\ dm^3$ have concentration $2\ mol/dm^3$.
1 mole of solute in $2\ dm^3$ has concentration $0.5\ mol/dm^3$.
0.5 mole of solute in $1\ dm^3$ has concentration $0.5\ mol/dm^3$.
... moles of solute in $2\ dm^3$ have concentration $1\ mol/dm^3$.
0.5 mole of solute in ... dm^3 has concentration $0.5\ mol/dm^3$.
0.1 mole of solute in $1\ dm^3$ has concentration ... mol/dm^3.
... mole of solute in $5\ dm^3$ has concentration $0.1\ mol/dm^3$.

By thinking about how to fill in the gaps, your pupils should be able to produce an expression for the number of moles in terms of the concentration and the volume:

$$\text{moles} = \text{concentration } (mol/dm^3) \times \text{volume } (dm^3)$$

It is important that your pupils are aware that the volume must be in dm^3 (or litres).

Examples

1. How many moles of potassium chloride are in $250\ cm^3$ of a $0.1\ mol/dm^3$ solution?

   $250\ cm^3 = 0.25\ dm^3$
   Moles = concentration × volume = $0.1 \times 0.25 = 0.025$

2. What is the concentration of a solution if 0.4 mole is dissolved and made up to $100\ cm^3$ of solution?

   $100\ cm^3 = 0.1\ dm^3$

   $$\text{Concentration} = \frac{\text{moles}}{\text{volume}} = \frac{0.4}{0.1} = 4\ mol/dm^3$$

3. What volume of solution will be needed if 0.05 mole of solute is to give a concentration of $2\ mol/dm^3$?

   $$\text{Volume} = \frac{\text{moles}}{\text{concentration}} = \frac{0.05}{2}$$

   $$= 0.025\ dm^3$$
   $$= 25\ cm^3$$

Preparation of a standard solution

A standard solution is a solution of known concentration. In practice, of course, solutions are made up by weighing out a certain mass of the solute and dissolving it to give a particular volume of solution. This means that calculations for practical purposes will have to involve moles = mass/RFM as well as moles = concentration × volume.

Example

What mass of barium chloride must be dissolved to make 5 dm³ of solution with a concentration of 0.5 mol/dm³?

Moles of $BaCl_2$ = volume × concentration
= 5 × 0.5
= 2.5

Mass of $BaCl_2$ = moles × RFM
= 2.5 × 208
= 520 g

Preparing a standard solution of sodium carbonate

The ability to make up a standard solution accurately is an important one. Pupils should plan to produce 250 cm³ of a solution of sodium carbonate, Na_2CO_3, of concentration 0.1 mol/dm³.

Materials
- watch glass
- spatula
- 100 or 250 cm³ beaker
- distilled water in wash bottle
- stirring rod
- filter funnel
- 250 cm³ volumetric flask with stopper
- anhydrous sodium carbonate, Na_2CO_3
- access to suitable balances (e.g. top pan, direct reading balance)

Safety
- *Anhydrous sodium carbonate is an irritant.*

Procedure
1. Pupils should calculate the mass of solute to be weighed out to make 250 cm^3 of solution.
2. Weigh out the calculated mass of solute on a watch glass. Alternatively, weigh the solute directly into the beaker.
3. Carefully transfer the solute to the beaker and, using distilled water from a wash bottle, rinse the watch glass into the beaker.
4. Add some more distilled water to the beaker and stir to dissolve the solute.
5. Using a funnel, transfer the solution to the volumetric flask.
6. Using distilled water, rinse the stirring rod, beaker and funnel into the flask.
7. Add distilled water to the volumetric flask to bring the solution up to the mark.
8. Stopper the flask and shake gently to mix the contents.

What mass of sodium carbonate should be used?

Moles of Na$_2$CO$_3$
= 0.1 × 0.25
= 0.025

RFM of Na$_2$CO$_3$
= (2 × 23) + 12 + (3 × 16)
= 106 g

Mass required to make 250 cm^3 of 0.1 mol/dm^3 solution
= 106 × 0.025
= 2.65 g

Possible problems
1. Overshooting the mark. This is usually prevented by carrying out the dissolving in a beaker that is much smaller than the volumetric flask.
2. Not fixing the stopper firmly and spilling some of the solution before it is properly mixed.

Calculating concentrations or volumes of solutions in reactions

This can be done in two ways. The first involves using moles, concentrations and volumes as appropriate for the substances in the question.

Method 1
What volume of 0.10 mol/dm^3 hydrochloric acid would be required to exactly neutralise 20 cm^3 of 0.05 mol/dm^3 aqueous sodium hydroxide?

$NaOH(aq) + HCl(aq) \rightarrow NaCl(aq) + H_2O(l)$

moles of NaOH = volume × concentration = $0.02 \times 0.05 = 0.001$

From the equation,

moles of HCl = moles of NaOH = 0.001

Hence,

$$\text{volume of HCl} = \frac{\text{moles}}{\text{concentration}} = \frac{0.001}{0.1} = 0.01 \text{ dm}^3 = 10 \text{ cm}^3$$

Method 2
The second method uses a general expression but it can only be used for one solution reacting with another.

$a A(aq) \quad + \quad b B(aq) \quad \rightarrow \quad c C(aq) \quad + \quad d D(aq)$

$$\frac{\text{volume of A} \times \text{concentration of A}}{\text{volume of B} \times \text{concentration of B}} = \frac{\text{moles of A}}{\text{moles of B}} = \frac{a}{b}$$

Example
What volume of 0.20 mol/dm^3 aqueous potassium hydroxide would be required to exactly neutralise 25 cm^3 of 0.05 mol/dm^3 sulphuric acid?

$2KOH(aq) + H_2SO_4(aq) \rightarrow K_2SO_4(aq) + 2H_2O(l)$

$$\frac{\text{volume of KOH} \times \text{concentration of KOH}}{\text{volume of H}_2\text{SO}_4 \times \text{concentration of H}_2\text{SO}_4} = \frac{\text{moles of KOH}}{\text{moles of H}_2\text{SO}_4} = \frac{2}{1}$$

Substituting, $\dfrac{\text{volume of KOH} \times 0.20}{25 \times 0.05} = \dfrac{2}{1}$

$$\text{volume of KOH} = \frac{2 \times 25 \times 0.05}{0.20} = 12.5 \text{ cm}^3$$

For this method it is not necessary to convert the volumes to cubic decimetres. The concentrations may be in g/dm^3 for *both* solutions.

5.7 Titration

◆ *Previous knowledge and experience*

Pupils will have experience of acids and alkalis and should appreciate that, if the correct amounts of acids and alkalis are mixed, a neutral solution will be produced. They will have met indicators such as litmus and Universal indicator.

They should be able to use a balanced symbol equation to calculate reacting masses and masses of products.

They should be able to add, subtract, multiply and divide simple numbers. They should understand simple ratios.

◆ *A teaching sequence*

Reactions between solutions are done as a titration when quantitative information is wanted. A standard solution is made up (or provided) and titrated against another solution of unknown concentration. From the volumes used of the two solutions and the concentration of the standard solution, which are known, the unknown concentration can be calculated.

Titration is the addition of one solution to the other in such a way as to stop at the point when the reaction is exactly completed. This *end-point* of the titration has to be made visible by using an indicator.

Performing a titration

Materials
- eye protection
- 2 beakers, 100 or 250 cm^3
- wash bottle
- pipette with safety filler
- 4 conical flasks, 100 or 250 cm^3
- burette
- burette stand
- filter funnel
- white tile
- litmus or pH paper
- suitable indicator, e.g. methyl orange or screened methyl orange
- hydrochloric acid solution (0.1 mol/dm^3)
- sodium hydroxide solution (0.1 mol/dm^3)
- Alternatively, pupils could use the standard solution of sodium carbonate previously prepared (see page 165).

!

<u>Safety</u>
- *0.1 mol/dm³ sodium hydroxide is an irritant.*
- *The use of the safety filler for the pipette and the tap of the burette needs practice.*
- *Anhydrous sodium carbonate is an irritant.*
- *Wear eye protection when using sodium hydroxide solution.*

<u>Procedure</u>
1. Rinse all the conical flasks with distilled water.
2. Using the safety filler, thoroughly wash out the pipette with the sodium hydroxide (or sodium carbonate) solution. (This is the alkali.)
3. Use the safety filler to fill the pipette to transfer a known volume of the alkali solution to each conical flask.
4. Add two or three drops of methyl orange or screened methyl orange to each flask.
5. Thoroughly wash out the burette with the hydrochloric acid solution.
6. Fill the burette with the hydrochloric acid solution and run out some of the solution to fill the jet.
7. Note the reading on the burette. (The reading does not have to be exactly zero.)
8. Add the solution from the burette to the alkali solution in the conical flask. Swirl constantly to mix, and frequently wash down the sides of the flask with distilled water. The end-point is reached when the indicator changes colour.
9. Note the reading on the burette and calculate the volume of hydrochloric acid used.
10. Repeat the titration three times, being very careful when approaching the end-point.
11. Calculate the average volume of hydrochloric acid solution used and then calculate the concentration of the 'unknown'.

<u>Possible problems</u>
1. Muddling up the two solutions. This can be prevented by having all the beakers carefully labelled, by keeping the pipette, safety filler, conical flasks and pipette solution on one side of the working area and the burette, funnel and burette solution on the other (as shown in Figure 5.3, overleaf). However, having litmus or pH paper handy allows for a quick check to be made if any doubt arises.
2. A bubble left underneath the burette tap. If this occurs, run out some solution quickly until the bubble is removed.
3. Not making sure there is enough solution in the burette before starting!

Figure 5.3
Front view of an arrangement of titration apparatus that helps to prevent the solutions becoming muddled.

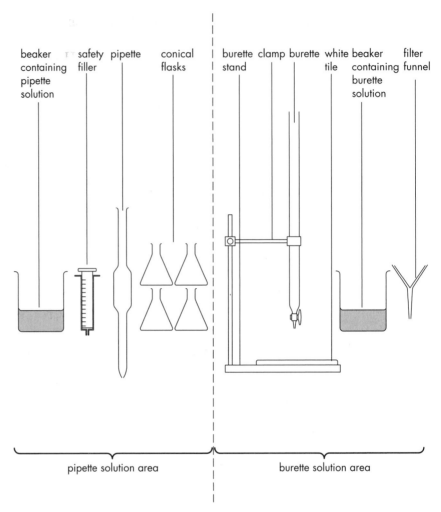

beaker containing pipette solution   safety filler   pipette   conical flasks   burette stand   clamp   burette   white tile   beaker containing burette solution   filter funnel

pipette solution area   burette solution area

What you might expect
The colour changes are:

- methyl orange: yellow in alkaline solution
  orange in neutral solution
  red in acid solution
- screened methyl orange (particularly useful if a pupil is colour blind)
  green in alkaline solution
  colourless in neutral solution
  purple in acid solution

The equations for the possible reactions are:

$$NaOH \quad + \quad HCl \quad \rightarrow \quad NaCl \quad + \quad H_2O$$

$$Na_2CO_3 \quad + \quad 2HCl \quad \rightarrow \quad 2NaCl \quad + \quad CO_2 \quad + \quad H_2O$$

The table below shows sample results for an experiment using 0.1 mol/dm^3 sodium carbonate solution and a solution of hydrochloric acid of unknown concentration.

Run	1*	2*	3	4*
final reading (cm³)	26.50	27.80	25.85	27.70
initial reading (cm³)	1.05	2.40	1.50	2.20
volume of acid used (cm³)	25.45	25.40	24.35	25.50

*Used in average.

Note that:

1. the burette is read to the nearest 0.05 cm^3
2. putting the final reading before the initial reading in the table makes the subtraction easier, and pupils will be less likely to make mistakes
3. in this case the third titration was not used as it would have led to an inaccurate average.

From these results, the average volume was 25.45 cm^3.

24.45 cm^3 of hydrochloric acid reacts with 25.0 cm^3 of 0.1 mol/dm^3 sodium carbonate solution.

25.0 cm^3 of sodium carbonate solution contains $0.025 \times 0.1 \text{ mol}$
$$= 0.0025 \text{ mol}$$

From the equation, 1 mole of Na_2CO_3 reacts with 2 moles of HCl.

Number of moles of HCl $= 2 \times 0.0025 = 0.005$

This is dissolved in 24.45 cm^3 of solution.

Concentration of HCl $= \dfrac{0.005}{0.02445} = 0.20 \text{ mol/dm}^3$

Suggested titrations

1. Make 250 cm³ of a 0.01 mol/dm³ solution of anhydrous sodium carbonate. Use it as a standard to find the concentration of a hydrochloric acid solution; use methyl orange as the indicator. (The acid used should have a concentration very close to 0.02 mol/dm³.)
2. Use a prepared solution of 0.005 mol/dm³ sulphuric acid as the standard to find the concentration of a sodium hydroxide solution; use methyl orange as the indicator. (The hydroxide used should have a concentration very close to 0.01 mol/dm³.)
3. Use a prepared solution of 0.01 mol/dm³ sodium hydroxide as the standard to find the concentration of an ethanoic acid solution; use phenolphthalein as the indicator. In this titration you may prefer to tell your pupils to put the alkali in the burette because it is easier to see the pink colour appearing, rather than disappearing. (The hydroxide used should have a concentration very close to 0.01 mol/dm³.)

Sample calculations

20 cm³ of 0.05 mol/dm³ potassium hydroxide was titrated with sulphuric acid four times. The first titration was a rough one, so the mean volume of acid used was found from the other three. It was 20.5 cm³. Find the concentration of the sulphuric acid.

$$2KOH \quad + \quad H_2SO_4 \quad \rightarrow \quad K_2SO_4 \quad + \quad 2H_2O$$
$$\text{2 moles} \qquad \quad \text{1 mole}$$

There are two alternative methods for the calculation:

Method 1 (using moles = volume × concentration)

Moles of KOH = volume (dm³) × concentration (mol/dm³)
$$= 0.02 \times 0.05$$
$$= 0.001$$

From the equation:

$$\text{Moles of } H_2SO_4 = \frac{\text{moles of KOH}}{2} = \frac{0.001}{2} = 0.0005$$

$$\text{Concentration of } H_2SO_4 = \frac{\text{moles of } H_2SO_4}{\text{volume (dm}^3)} = \frac{0.0005}{0.0205}$$

$$= 0.024 \text{ mol/dm}^3$$

Method 2 (using the general titration formula)

$$\frac{\text{volume of } H_2SO_4 \times \text{concentration of } H_2SO_4}{\text{volume of KOH} \times \text{concentration of KOH}} = \frac{\text{moles of } H_2SO_4}{\text{moles of KOH}} = \frac{1}{2}$$

Substituting,

$$\frac{20.5 \times \text{concentration of } H_2SO_4}{20.0 \times 0.05} = \frac{1}{2}$$

$$\text{Concentration of } H_2SO_4 = \frac{1 \times 20.0 \times 0.05}{2 \times 20.5} = 0.024 \text{ mol/dm}^3$$

Calculations from reactions involving solids and solutions

For reactions that involve a solid and a solution, both moles = mass/RFM and moles = volume × concentration will be needed.

Examples

1.   What volume of 2.0 mol/dm³ hydrochloric acid will be needed to react completely with 8.0 g of copper oxide?

$$CuO + 2HCl \rightarrow CuCl_2 + H_2O$$

$$\text{Moles of CuO} = \frac{\text{mass}}{\text{RFM}} = \frac{8.0}{80} = 0.1$$

From the equation,

moles of HCl = moles of CuO × 2 = 0.1 × 2 = 0.2

Hence,

$$\text{volume of HCl} = \frac{\text{moles of HCl}}{\text{concentration}} = \frac{0.2}{2.0} = 0.1 \text{ dm}^3 = 100 \text{ cm}^3$$

2.   What mass of calcium nitrate would be formed if 10 cm³ of 0.1 mol/dm³ nitric acid was reacted with excess calcium carbonate?

$$CaCO_3 + 2HNO_3 \rightarrow Ca(NO_3)_2 + CO_2 + H_2O$$

Moles of HNO_3 = volume × concentration = 0.01 × 0.1 = 0.001

From the equation,

$$\text{moles of } Ca(NO_3)_2 = \frac{\text{moles of } HNO_3}{2} = \frac{0.001}{2} = 0.0005$$

Hence,

$$\text{mass of } Ca(NO_3)_2 = \text{moles of } Ca(NO_3)_2 \times \text{RFM}$$
$$= 0.0005 \times 164 = 0.082 \text{ g}$$

◆ *Enhancement ideas*

- ◆ Vinegar is a dilute solution of ethanoic acid. Plan and carry out an experiment to find which of a selection of shop vinegars is the best value. (Possible suggestion: titrate each of them against a known volume of a standard alkali and see which vinegar is needed in the smallest volume.)
- ◆ Washing soda (hydrated sodium carbonate) loses its water of crystallisation when left open to the air. Plan and carry out an experiment to find the value of x in $Na_2CO_3.xH_2O$, for an old sample of washing soda. (Possible suggestions: use the direct 'water of crystallisation' method or titration – dissolve a known mass in a known volume, titrate it against standard acid, find mass of sodium carbonate in the sample, subtract to find mass of water and calculate x.)
- ◆ Indigestion tablets react with hydrochloric acid in the stomach. Plan and carry out an experiment to find which of a selection of indigestion tablets is the best. (Possible suggestion: find which tablet neutralises the greatest volume of acid.)
- ◆ A metal oxide, formula MO, reacts with acid. Plan and carry out an experiment to find the relative atomic mass of M and then to suggest its name. (Possible suggestion: find the volume of standard acid neutralised by a known mass of the oxide. Calculate the number of moles of the oxide, then the RFM, and hence find the RAM of metal M.)

Two examples to challenge your more able pupils
- ◆ $150\,cm^3$ of $0.5\,mol/dm^3$ hydrochloric acid was added to $3.1\,g$ of copper carbonate. Calculate **a)** the mass of salt formed, **b)** the volume of gas given off, at r.t.p., **c)** the volume of $0.25\,mol/dm^3$ sodium hydroxide needed to neutralise the remaining acid.
- ◆ $10\,cm^3$ of a hydrocarbon C_xH_{2y} burns in excess oxygen to give $30\,cm^3$ of carbon dioxide and $40\,cm^3$ of water vapour (all measurements being made at the same temperature and pressure). Use Avogadro's law to find the formula of the hydrocarbon.

◆ *Other resources*

Web sites

◆ This section of the syllabus is covered well in **Chemistry School.com**, a combined CD ROM and internet site developed by the Nuffield Foundation to support KS3 and KS4 Chemistry. It is available from New Media Press, PO Box 4441, Henley-on-Thames RG9 3YR.

Video

A video and informative teachers' notes on the mole can be purchased from Classroom Video at Darby House, Bletchingley Road, Merstham, Redhill, Surrey RH1 3DN.

Background reading

One problem you may find with this topic is gathering enough examples of calculations for your pupils to use. The following books should help.

Brown, Keith (1986). *Moles, a survival guide for GCSE Chemistry.* Cambridge University Press. ISBN 0 5214 2409 7.

Ramsden, E.N. (1994). *Calculations for GCSE Chemistry.* Stanley Thornes. ISBN 0 7487 1738 2.

Hunt, J.A. & Sykes, A. (1985). *Chemistry calculations.* Addison Wesley Longman. ISBN 0 5823 3181 1.

Latest nomenclature can be found in *Signs, symbols and systematics* published by the Association for Science Education (1995), College Lane, Hatfield, Hertfordshire.

6 Geological changes

John Payne

6.1 Weathering and erosion
Neighbourhood walk
Expansion of water on freezing (the
  freeze–thaw mechanism)
The effect of extremes of heat and cold
Which rocks are attacked by acid rain?
Modelling an erosion process

6.2 Sedimentary rocks
Sedimentation
Modelling particle size and compaction
Experiment to examine deposited sediments
Making fossils

6.3 Other types of rock: igneous and metamorphic
Igneous rocks
The effect of cooling rate on crystal size
Basalt and granite
Metamorphic rocks
Demonstration of the formation of slate

6.4 The rock cycle
Recycling
The rock cycle
Rock samples

6.5 Plate tectonics
Convection currents in the mantle
Continental crust and oceanic crust
Evidence for plate tectonics
Plate boundaries
Folding and faulting

♦ *Choosing a route*

In this section we shall be looking at erosion of rocks, rock types, the rock cycle, the ideas behind plate tectonics and the consequences of plate movement.

Although it has now been part of the National Curriculum for many years, the geology section of Science syllabuses still seems to be approached with some dread by some science teachers. This can be easily understood since, even for younger teachers, it was not part of 'our' science curriculum at school. In many ways, therefore, it is a new subject area whether we be biologist, chemist or physicist. One useful source of expertise in schools has often been the Geography department, whose province geology once was in its entirety. However, the introduction of this topic into the Science department has brought with it the chance to use laboratory facilities for practical work. The 'science' approach is much more concerned with processes and their causes, rather than a description of the resulting land-forms and their relationship to people, which Geography still covers.

Since the formation and subsequent weathering of rocks is part of the rock cycle, it would be possible to have a concept map also in the form of a cycle. It would actually be possible to start anywhere within this cycle. However, many pupils find it easier to start with something they can easily see, and so it is suggested that erosion and weathering are looked at first and the concepts developed as shown. Although many textbooks will also follow this route, some may begin with tectonic ideas in order to explain the rock cycle as a whole.

6.1 Weathering and erosion

♦ *Previous knowledge and experience*

Pupils may have taken part in a 'town trail' exercise and looked for examples to show that the weather can wear rocks away. Some may already appreciate that besides natural weathering there is such a thing as acid rain which is wearing away some rocks even faster. If pupils have done a primary school project, they may have looked at erosion by rivers, the wind, the sea and by glaciers.

♦ *A teaching sequence*

The processes of weathering and erosion are frequently confused. Weathering is the process taking place when gases and water in the atmosphere combine with surface water and solar radiation to break up surface rocks. Weathering can be of three main types:

- Mechanical weathering (sometimes known as 'physical weathering'). During the day rocks heat up and expand. At night rapid cooling causes stresses. This happening over and over again breaks down rocks. Another mechanism is 'freeze–thaw' action. Water expands when it freezes, so water in cracks expands when the water turns to ice. This breaks down rocks. The effect of these two very slow processes is to reduce the size of the rock fragments and increase the surface area on which chemical action can take place.
- Chemical weathering. This is the breakdown of rocks by chemical reaction. An example is the breaking down of limestone with rain water. The rain water contains dissolved carbon dioxide and forms carbonic acid:

$$H_2O + CO_2 \rightarrow H_2CO_3$$

This attacks the limestone:

$$CaCO_3 + H_2CO_3 \rightarrow Ca(HCO_3)_2$$

- Biological weathering. As plant roots grow, they increase in girth and this produces stresses on rocks, which break up.

Processes of weathering are relatively simple to understand.
Erosion is the breaking down of rocks by movement of rivers, ice, sea or wind. This topic needs to be linked to the study of energy transfer in Physics.

Neighbourhood walk

A walk in the local neighbourhood is a good way to start the topic. Some pupils may have done this at the primary stage but it is still good to reinforce the 'out in the field' aspect of geology. What is actually observed will depend on the locality, but in any location look for examples of where rocks, or products made from rocks, have been worn away. The problem of comparing the amount of weathering with the time the rock has been exposed to the weather should present an interesting discussion.

Hopefully someone will suggest the idea of using grave stones, since these are clearly dated. The engraved lettering on older grave stones should be more worn away and less easy to read. There are some good opportunities here for fair testing:

- Is the material of the grave stones identical?
- Is the size and depth of the engraving similar?
- Are all the grave stones equally exposed to the prevailing wind and rain?

More able pupils may be able to link the type of rock to the rate at which it is worn away, granite (igneous) and marble (metamorphic) being much harder than either limestone or sandstone (both of which are sedimentary). Ask what processes cause the grave stones to be worn away. The mechanical action of rain and wind-blown solid particles may be suggested, but it would be nice if the action of frost, extremes of temperature and, for some rocks, chemical action of acid rain came from the discussion.

A map of the proposed trail is useful and a prior visit by the teacher taking the group essential. Any set procedures for taking pupils out of school must also be followed. Ask permission to visit graveyards in advance and remind the pupils to treat the area with respect.

Expansion of water on freezing (the freeze–thaw mechanism)

Since setting up this experiment requires very little time, it may be possible to do it after the 'town trail' walk. The discussion about the results can take place when there is something to see.

Materials
- small glass bottle with screw top
- small plastic bottle with screw top
- water
- plastic bags to hold the bottles
- access to freezer

<u>Safety</u>
- *Warn anyone likely to use the freezer of the danger of broken glass fragments or put a warning label on the freezer.*

<u>Procedure</u>
1. Fill the glass bottle and the plastic bottle right to the top with water. Screw their tops on well.
2. Seal both bottles in plastic bags.
3. Place in the freezer and leave overnight.

Figure 6.1
Modelling the freeze–thaw mechanism with water in a glass bottle.

plastic bag

glass bottle of water

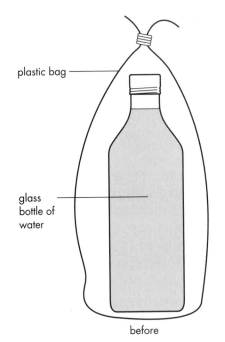
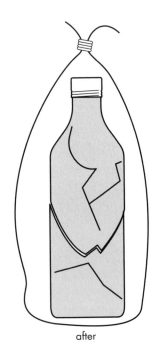

before   after

<u>What you might expect</u>
Water is unusual in that it expands when it freezes. The glass cannot expand and cracks (Figure 6.1). Notice how the cap may have been bent slightly outwards too. Link this to the idea of water in rock cracks freezing and opening up the cracks. More able pupils may comment on whether the practical is an accurate representation of what is meant to be being modelled – 'Surely there is no lid on the ice forming in the crack?' The best answer is that the water on the surface freezes first, forming a lid; the water lower down freezes later and cannot then expand upwards.

A possible extension is to repeat this experiment with various sorts of plastic bottles, which are difficult to crack even when the plastic seems to be quite rigid.

Links can be made here to metal pipes bursting in winter and milk freezing on doorsteps in glass bottles. The bottle does not crack this time: the soft top is pushed up instead.

The effect of extremes of heat and cold

<u>Materials</u>
- eye protection
- glass rod
- Bunsen burner
- beaker of cold water
- tongs

<u>Safety</u>
- *This experiment involves hot materials and breaking glass.*
- *Wear eye protection when using a Bunsen burner, and especially placing the hot glass in the cold water.*

<u>Procedure</u>
1. Heat the glass rod.
2. Plunge it into the beaker of cold water.

Figure 6.2
Modelling the effect of extreme temperatures.

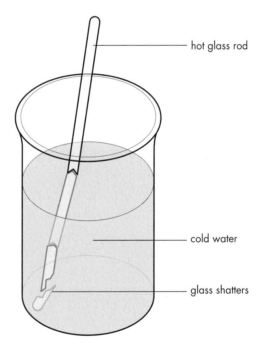

hot glass rod

cold water

glass shatters

<u>What you might expect</u>
The piece of glass breaks when it is plunged into the cold water (Figure 6.2). This is due to the tremendous pressures which exist in glass when it is cooled rapidly. Point out to pupils that what we have done with the glass gives a much greater difference in temperature than rocks would experience in nature, although the results are the same; we have simply speeded up the process. The so-called 'onion skin' effect is when rounded boulders lose their outer skin by this effect and layers of rock peel away.

Which rocks are attacked by acid rain?

Some rocks are weathered quickly by rain that contains acids. Point out to pupils that although they will be using dilute hydrochloric acid to represent 'acid rain', this is not what is actually present. Do not be tempted to use dilute sulphuric acid in this experiment, since the reaction between this and calcium carbonate produces the insoluble salt calcium sulphate and so may not give such obvious results.

Materials
- eye protection
- small samples of a range of rock types, e.g. limestone, marble and chalk (all of which will give positive results) and at least three others, e.g. granite, gneiss and slate
- bottle of hydrochloric acid (0.1 mol/dm³)
- watch glasses
- dropping pipettes

Safety
- *Be careful when cleaning up, as the watch glasses may still contain unreacted hydrochloric acid.*
- *Wear eye protection when using hydrochloric acid.*

Procedure
Put a few drops of the acid on to each rock in a watch glass (Figure 6.3) and look for evidence of a chemical reaction (fizzing).

Figure 6.3
Using dilute hydrochloric acid to model attack by acid rain.

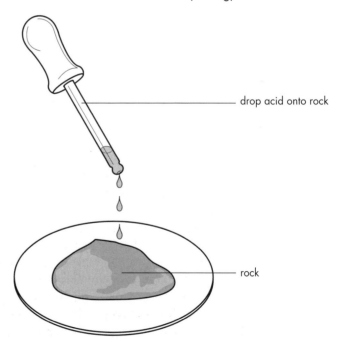

drop acid onto rock

rock

<u>What you might expect</u>
There should be fizzing with limestone, marble and chalk. Granite, gneiss and slate have no immediate reaction. More able pupils may point out that the granite is affected by acids and forms clay. If so, point out to them that this happens over a very long time period.

Point out that the acid being used is stronger than that present in acid rain and that we have speeded up the process to get quickly observable results.

Link the results to why some statues and old buildings have eroded faster than others, especially in the last hundred years. This provides an opportunity for a scientific investigation into the way that the strength of the acid in acid rain would affect how quickly limestone or marble is weathered (see Chapter 8).

Modelling an erosion process
When pieces of rock move about and collide with each other, they wear each other away. This leads to smaller pieces without sharp edges. In this activity pieces of brick are used rather than rock, as the effects of erosion are seen more quickly.

<u>Materials</u>
- 21 pieces of brick (similar sizes 5–7 mm)
- wide-necked plastic bottle with a screw top
- colander
- tray to collect fine residue
- access to balance

<u>Safety</u>
- *Use a well-ventilated laboratory.*

<u>Procedure</u>
1. Keep one piece of brick for comparison later.
2. Weigh the 20 pieces of brick.
3. Put the 20 pieces of brick into the plastic container and screw the lid on tightly.
4. Shake the container vigorously for exactly 1 minute (see Figure 6.4).
5. Empty the container into the colander. Select all the pieces greater than 5 mm across. Count these and weigh them.
6. Put all the pieces larger than 5 mm back into the plastic container.
7. Repeat steps 4 to 6 as many times as you can.
8. Plot the results in a graph (see below).
9. Compare the pieces left with the original piece kept for comparison.

Figure 6.4
*Modelling erosion
with brick pieces
in a plastic bottle.*

<u>What you might expect</u>
The pieces of brick break down during the experiment and their rough
edges are removed. Pupils can plot different graphs:

- the number of pieces against the number of times the container had
  been shaken
- the mass of the lumps remaining against the number of times the
  container had been shaken.

♦ *Enhancement ideas*

- ♦ Posters can be made to produce a class display. These
  posters could distinguish different methods of weathering
  and erosion.
- ♦ The experiment involving shaking pieces of brick could
  be repeated using different rocks. Pupils could make
  predictions of which rocks will be eroded the most.
- ♦ Repeat by heating a range of rock samples to high
  temperature followed by cooling in a deep freezer.

6.2 Sedimentary rocks

◆ *Previous knowledge and experience*

More able pupils may have come across this term and may even know the features of sedimentary rocks and how they were formed. To many, however, the term will be new but it can easily be linked to the idea of sediments settling.

◆ *A teaching sequence*

The key idea here is to link the transport and subsequent accumulation of the small solid particles of weathered rock with the formation of new rock. It is important to emphasise that the time scale of the process is millions of years. Pupils should realise that these particles have to be transported, usually by water, wind or glaciers.

Sedimentation

The particles settle as sediments. Older layers are always lower down. Over millions of years, the oldest layers of sediment come under great *pressure* from material above them. This squeezes out the water but leaves behind minerals that were dissolved in the water. These minerals help *cement* the particles together to form a sedimentary rock. This can be modelled in the laboratory.

Modelling the sedimentation process
Materials
- eye protection
- plastic syringe
- dry sand
- powdered clay
- plaster of Paris
- petroleum jelly
- file
- variety of masses

Safety
- *Do not put plaster of Paris down the sink.*
- *Wear eye protection when using plaster of Paris.*

GEOLOGICAL CHANGES

Figure 6.5
Modelling the formation of sedimentary rocks.

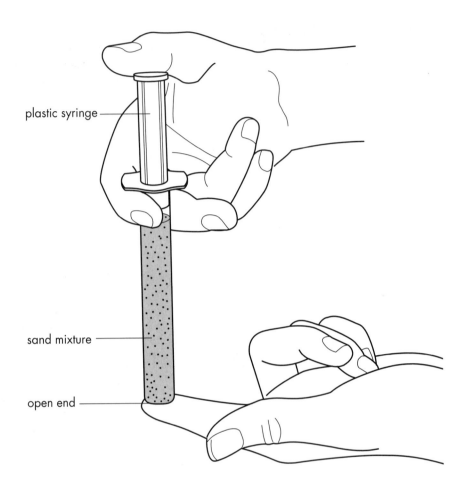

plastic syringe

sand mixture

open end

<u>Procedure</u>
1. Cut off the end of the syringe so that it is an open barrel.
2. Smear petroleum jelly on the inside of the syringe.
3. Put four measures of damp sand inside the syringe. (Adjust the actual quantities to suit the size of your syringe; a large spatula-full should be about right for one measure.)
4. Block the open end of the syringe with a finger or thumb and press the plunger down as hard as possible (see Figure 6.5).
5. Stop pressing, remove the finger and carefully push out the sand pellet and leave it to stand on a piece of paper to dry.
6. Mix three measures of damp sand with one measure of clay and repeat steps 3, 4, and 5.
7. Mix five measures of damp sand with one measure of plaster of Paris and repeat steps 3, 4 and 5.
8. Leave the 'rocks' to dry overnight.

What you might expect

The least 'rock-like' pellet should be the one with just the damp sand; this is equivalent to pressure alone and no cementing agent. The most 'rock-like' will be the one with plaster of Paris acting as cementing agent.

Pupils can then devise tests to compare the 'strengths' of the 'rocks'. They could use a file on them or drop increasingly large masses on them from a fixed height. (Be aware of safety if pupils are dropping large weights.) The idea to reinforce is that pressure on its own has not produced a very strong rock, but the presence of cementing agents has.

Modelling particle size and compaction

Pupils may be able to appreciate that the smaller the particles of sediment, the greater the amount of water that can be squeezed out and so the greater the compaction. It is possible to model this using large beads or marbles and small ones. Put the large beads (representing the large particles, such as sand grains) on an overhead projector and push them together with two wooden blocks; not much compaction is possible. Now do the same with smaller beads (representing smaller particles, such as mud particles); much more compaction is possible showing that more water could be squeezed out.

Reinforce the idea that sand grains become sandstone when cement crystallises in the spaces between the particles and glues the particles together. The cement is formed from minerals that crystallise from the solution that was once between the particles. Pressure to cause this comes from the weight of the sediments above.

Experiment to examine deposited sediments

Materials
- large glass beaker or jar
- 600 cm^3 beaker
- soil

Safety
- *Pupils should wash their hands after handling soil.*

Procedure
1. Half-fill the large beaker with water.
2. Sprinkle soil into it from the smaller beaker until there is a layer 1 cm thick at the bottom.
3. Leave the container for 2 days.
4. Add another layer of soil and leave for a further 2 days.
5. Finally add one more layer and allow to settle for 2 further days.

<u>What you might expect</u>
Pupils should notice some degree of sorting taking place: the larger fragments become concentrated at the base of each layer with the finer fragments on top. They will notice that coarse particles sink faster than finer ones.

When a river is running into the sea, the largest particles will settle out close to the shore, and the finer particles are carried out further. You can show pupils a piece of conglomerate which would form close to the mouth of the river. The oldest layer is the one at the bottom and the youngest is on top.

Making fossils

Fossils are found in many sedimentary rocks but never in igneous rocks. They may be found, often very distorted, in metamorphic rocks that have been formed by the action of pressure alone. It is probably best for pupils to associate fossils with sedimentary rocks and appreciate the use that fossils can have in dating sedimentary rocks. Fossils can be formed in four ways:

- preservation of the whole fossil, e.g. woolly mammoth
- petrification (turning into rock), e.g. petrified forests
- impression, e.g. footprints
- casting.

Modelling fossil formation

It is possible to model the formation of a fossil by casting in sedimentary rocks.

<u>Materials</u>
- eye protection
- Plasticine
- selection of shells (ones with pronounced external ridges are best)
- plastic disposable cup
- hand lens
- plaster of Paris

<u>Safety</u>
- *Do not put plaster of Paris down the sink.*
- *Wear eye protection when using plaster of Paris.*

<u>Procedure</u>
1. Soften the Plasticine.
2. Make a mould by pressing the shell into the Plasticine. It is important to keep enough plasticine around the edge of the impression to form a rim. Pupils should push the shell in and then lift it out vertically, avoiding any sideways movement which would destroy the shell impression.
3. Mix up some plaster of Paris in a disposable cup: add just enough water, with stirring, to make a runny cream.
4. Pour the cream into the mould. (It may be best, if disposable cups have been used, to let any excess plaster of Paris dry in the cups.)
5. When the casts have set, their detail can be compared, using hand lenses, with that of the original shells.

<u>What you might expect</u>
Pupils should get close replicas of the original shells with good detail. It is important now for pupils to grasp the idea of casting and to appreciate that the plaster of Paris cast is not the fossil itself. What happens is that an animal may die and become trapped as layers of sediment form around it. Eventually, the remains of the organism decay but the space it occupied is still there. This space can be filled with minerals that crystallise and form a cast of the original shape of the organism that died. What we then call the fossil is just like our plaster impression – it is a copy of what originally occupied the space.

◆ *Enhancement ideas*

- ◆ A study could be undertaken of fossils and the geological time periods that they typify. Some species have become extinct and we only know about them from the fossil record. Species alive today have ancestors stretching back millions of years.

- ◆ Examination of typical sedimentary rocks could be left until later, when a comparison with other rock types can be made.

6.3 Other types of rock: igneous and metamorphic

◆ *Previous knowledge and experience*

More able pupils may have encountered the words 'igneous' and 'metamorphic'. They may be able to link the word igneous to rocks that have formed or come from lava or volcanoes. It is less likely that they will have come across the term 'metamorphic', although they may have seen 'Morph' on the television!

◆ *A teaching sequence*

Pupils have so far been introduced to erosion and the subsequent transport of eroded material. This material may possibly form sedimentary rock over millions of years. At this stage they may also have come across the names of the other two rock types but without any great detail about their formation or characteristics. Development of ideas about igneous and metamorphic rock types will inevitably lead towards the concept of the rock cycle as a whole.

Igneous rocks

Pupils first need to appreciate that the Earth has a solid crust which is floating on very viscous molten rock called the 'mantle'. The thickness of this crust ranges from between 5 and 10 km at its thinnest (under the oceans), to between 25 and 40 km in continental crust. Show them a hard boiled egg and ask which is relatively thicker, the egg shell compared with the egg as a whole, or the Earth's crust compared with the Earth as a whole. The answer is that the Earth's crust is relatively much thinner. Given the thickness of the Earth's crust and the Earth's diameter, more able pupils may be able to use a micrometer to measure the thickness of an egg shell and the diameter of the egg and do the necessary mathematics to demonstrate the relative thicknesses of each.

The next step is to point out that the molten rock in the mantle can exploit weaknesses in the crust and rise towards the surface. It may emerge through the surface as volcanoes, forming extrusive igneous rocks, or it may not reach the surface but cool underground, forming intrusive igneous rocks.

This movement of molten rock towards the surface can be modelled.

Modelling the flow of magma through the crust

Materials
- eye protection
- 250 cm³ beakers
- red candles (or red candle wax)
- sand
- water
- Bunsen burner
- tripod, gauze and heat proof mat

Safety
- *Wear eye protection when using a Bunsen burner and when looking closely into the hot beaker.*

Procedure
1. In advance of the class experiment, the beakers and their contents must be assembled. Melt the wax and pour about 10 cm³ of it into the bottom of each beaker. Allow the wax to set. Cover it with about 15 cm³ of sand and then add 25 cm³ of water.
2. Give the pupils the beakers and tell them to warm the bottom of the beaker very gently. It may be best to warm just one spot on the beaker.
3. Watch carefully through the side of the beaker.

What you might expect

Although this may seem a very simple experiment, pupils do seem to enjoy it. It does, however, need quite a conceptual leap to appreciate how it models magma rising through layers of crust. First, pupils should be told that both the sand *and the water* should be imagined as layers of *solid* crust. As the wax melts, it becomes less dense and rises through the layers of crust to reach the surface. The lower density of magma is the reason why it rises towards the surface of the Earth. There it flows out as lava and cools.

The class may get a complete variety of features and results from their beakers (Figure 6.6, overleaf). It often seems that all of the wax passes to the surface in just one or two very thin tubes (similar to volcanoes). These are above the hot spot where the Bunsen burner was placed (similar to hot spots in the mantle). If the heating is very gentle, and stopped before all the wax has reached the surface, it may be possible to see magma chambers forming in the water (modelling the fact that some magma rises but does not reach the surface, staying trapped underground and forming intrusive igneous rock).

It is not possible to model a weakness in the 'crust' to show where the magma rises, although the fact that it rises in thin tubes must indicate that some weakness in the sand layer has been exploited.

Figure 6.6
Using candle wax, sand and water to model the movement of molten rock beneath the Earth's surface.

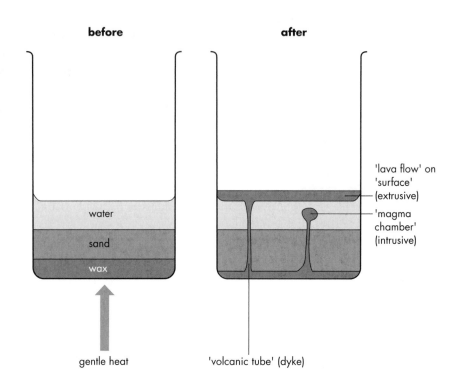

The effect of cooling rate on crystal size

Any rock that is formed from molten rock cooling is called an *igneous rock*. Molten rock is called *magma* until it emerges on to the surface, then it is *lava*. The rate at which molten rock cools influences the size of crystals that form in the resulting igneous rock. Given lots of time to cool, large crystals will form. Rapid cooling produces lots of small crystals.

Materials
- salol (phenyl salicylate)
- 250 cm³ beaker (used as a water bath)
- test-tube
- thermometer
- dropping pipette
- tripod, gauze and heatproof mat
- microscope slides
- overhead projector (optional)
- hand lenses
- samples of basalt and granite
- access to freezer and oven

<u>Procedure</u>

1. Before the start of the experiment put some microscope slides in the freezer and another batch of slides in a warm oven (40–45°C).

2. In the 250 cm^3 beaker warm about 100 cm^3 of water to about 55°C.

3. Put two measures of salol into a test-tube. Put the test-tube in the beaker of warm water; the salol should melt.

4. Put a dropping pipette in the molten salol, and allow the end of the pipette to reach the same temperature as the molten salol.

5. Get one of the cold microscope slides. Put it on the bench wedged at a slight angle (e.g. on a pencil, see Figure 6.7).

Figure 6.7
Cooling molten salol at different rates.

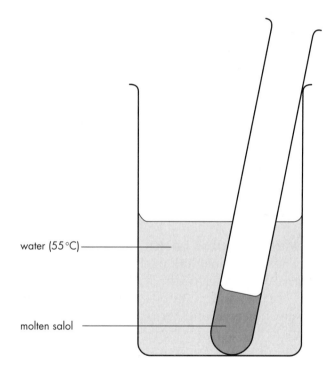

water (55°C)

molten salol

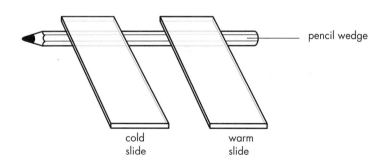

pencil wedge

cold slide   warm slide

6. Quickly transfer a few drops of the molten salol to the top of the slide and allow them to run down.
7. Now do the same on a warm slide (if crystals do not start to appear, try dropping a small 'seed' crystal into the melt).
8. Use hand lenses to look at the crystals.
9. It may then be possible to do this as a demonstration, with the slides on an overhead projector.

<u>What you might expect</u>
Results from this experiment are usually reliable, but not always. It may be that some good examples are found and passed around the rest of the class. If the salol cools quickly there are lots of small crystals. If it has the chance to cool more slowly, larger ones will form.

Basalt and granite
Now is a good time to look at samples of basalt and granite.

Basalt has small crystals, in reality so small they are hard to see with the naked eye. This is an example of an extrusive rock that has cooled quickly from molten material.

Granite has clearly visible, larger crystals; they have formed underground from magma that did not reach the surface. The molten material had lots of time to cool down slowly, so larger crystals formed.

Metamorphic rocks
This is the final link in the three rock types, which will then enable the idea of the rock cycle as a whole to be presented.

At its simplest, pupils need to understand that some rocks may not melt and form part of the mantle but can be affected by heat and/or pressure in such a way that their appearance and physical properties are altered. The minerals making up the rock have their structures changed without actually melting ('metamorphosed' means 'changed').

Metamorphism usually occurs deep underground (regional metamorphism) but can also occur close to volcanic tubes. The rock near the magma gets baked and can metamorphose. This is called 'contact metamorphism' and can be modelled in a very simple way.

Modelling metamorphism
<u>Materials</u>
• clay
• access to an oven

Procedure
1. Roll up some clay into balls of various sizes (2 cm to 5 cm in diameter).
2. Bake the clay balls in an oven at approximately 105 °C.
3. Cool, then examine.

What you might expect
The baked clay can be compared with samples of the original clay. Its appearance has changed (it has become lighter in colour) and it has become much harder. It has not burned in the sense of reaction with oxygen (this could not happen underground), and none of the minerals in the clay have been lost. However, the substances present in the original clay may have had their structures changed by the heat and the new structures formed have different properties. Breaking into the larger clay balls, the centres should be found to be little changed or unchanged (unless excessive baking has been done); this models the effect of being further away from the heat that is causing the metamorphic change. The idea can also be modelled by using the idea of baking a fruit cake.

The main point to bring out is the changed appearance (lighter colour) and properties (harder). Do emphasise that this experiment is only a very approximate model of contact metamorphism. Pupils should imagine the effect that the magma in volcanic tubes will have on the rock that is close to it.

For regional metamorphism, it is important to bring out the idea that high pressure (possibly with heat too) can also change the appearance and properties of rocks and that the greater the pressure or heat, the greater the metamorphic effects will be.

Demonstration of the formation of slate
This experiment models the formation of slate (metamorphic) from mudstone (sedimentary) and models how properties change (in this case, cleavage).

Materials
- overhead projector
- matchsticks
- 2 blocks of wood
- kitchen knife or similar
- samples of slate (enough to cleave)
- samples of schist and gneiss
- hand lenses

Figure 6.8
Using matchsticks to model the properties of mudstone and slate.
a Random arrangement of particles in mudstone.
b Alignment of the particles to form slate.
c The 'slate' will now cleave easily parallel to the planes of particles.

a

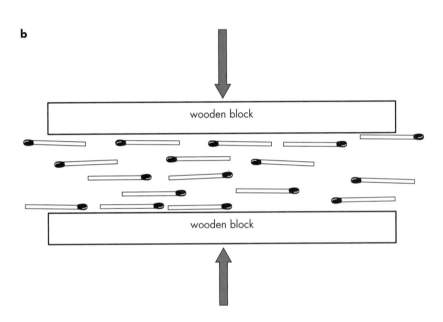

b

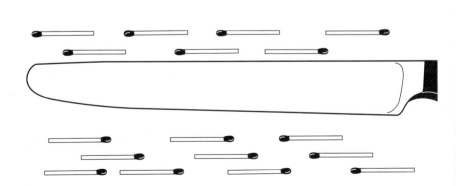

c

Procedure
1. Put the matchsticks fairly randomly on the overhead projector; these represent the particles in mudstone (see Figure 6.8a).
2. Try to push the knife through the centre of the matchsticks: the particles just scatter about.
3. Now push the matchsticks as close together as possible, using the wooden blocks to represent the effect of pressure: the particles now line up in rows (see Figure 6.8b).
4. Remove the blocks. Now push the knife through in the direction of the plane of the particles (see Figure 6.8c). The particles move apart, leaving smooth surfaces. This is called 'cleaving'.
5. Try cleaving the assembly of matchsticks in any other direction. It will not cleave and is very resistant indeed. This allows the idea to be developed that the direction of the force causing the metamorphism is at right angles to the layers of particles. (Refer back to the direction the wooden blocks were pushed.)
6. Now demonstrate or get the pupils to cleave samples of real slate. (It may be best to get slate samples that can be easily cleaved to avoid the use of hammers, chisels, etc.)

What you might expect
This is a model. More able pupils may realise that particles in minerals are not 'long and thin' like the matchsticks. Admit this, but explain that it is the idea that is important.

Slate is formed under relatively mild conditions. Indeed, some fossils can still be present in slate although they can often be very distorted. More extensive metamorphism results in the different minerals in the rock recrystallising in bands or layers. Samples of schist or gneiss can be used to show this. Gneiss is the most heavily metamorphosed and bands of different minerals can be seen in it. There is no easy way to model the formation of this in the laboratory, but a close look at samples of gneiss can bring out important facts.

Minerals present in the gneiss may include quartz (usually colourless or white), feldspar (colourless or white, sometimes pink), mica (could appear grey–brown or black) and garnet (red, brown, grey, yellow or black!). More able pupils may be able to research the identities of the minerals. However, the main point to bring out is that all of these minerals, which are now so clearly visible, were present in the original clay or mudstone; it is impossible to see individual minerals in these. The clay or mudstone has metamorphosed to slate and then to gneiss. In the gneiss the minerals have accumulated in layers of their own type (layers of quartz, layers of feldspar and so on). The direction of these layers can give evidence of the direction of the forces that caused the metamorphism.

◆ *Enhancement ideas*

◆ The fossils that can be found in slates are often very distorted. The distorting forces that acted on the rock can be deduced from the way that a fossil cast is itself distorted (e.g. squashed from side to side, top to bottom, or some sort of sheering force). The experiment on page 188 can be repeated but the mould distorted by pressure after the impression has been made but before pouring in the plaster of Paris. Once the cast has set, it should be possible to deduce what forces must have acted to cause the distortion. This models what geologists could do to deduce the sizes and directions of distorting pressure in the formation of metamorphic rocks.

◆ Although the experiment on page 192 models igneous rocks cooling, real examples of igneous rocks contain more than one mineral. These different minerals all cool at different rates and may crystallise at different times in the cooling process. More able pupils could research what minerals are present in granite samples. They could then compare the average crystal sizes of each of the different minerals. The main minerals in granite samples will be quartz, feldspar and mica.

◆ Examination of samples of 'ropey' lava can provide for interesting discussion on how this rock formed. The clear threads of rock on the surface show the direction of the lava flow as it cooled quickly on the surface, leaving the pattern in place. Underneath, the porous looking nature of the rock shows that it cooled quickly enough to trap bubbles of air under the surface.

6.4 The rock cycle

◆ *Previous knowledge and experience*

It is likely that only the most able pupils, those who have done extended project work on rocks or those with a very keen interest, will have met the concept of the recycling of rocks.

◆ *A teaching sequence*

Recycling

Pupils will be familiar with recycling and will be able to give examples of how paper, metals and glass are recycled. The rock cycle is best approached following a discussion of recycling.

Figure 6.9 shows the steps in recycling a glass bottle to produce a new bottle. You could show the cycle overhead on an overhead transparency, with the labels for the arrows missing. These labels could be on separate pieces of acetate and, in discussion, pupils could decide which label goes with which arrow.

Figure 6.9
The recycling of glass. The words in italics represent the equivalent processes in the rock cycle. (Source: World of Science 2 *by G. Booth, B. Mc Duell & J. Sears, 1999, by permission of Oxford University Press.)*

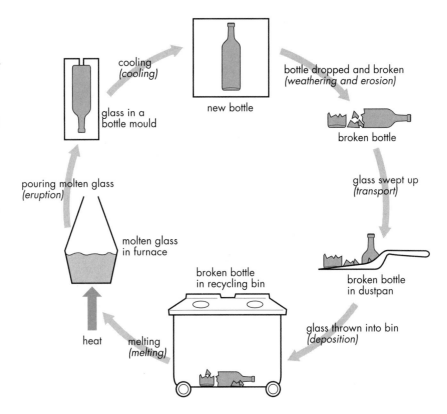

cooling
(cooling)

new bottle

bottle dropped and broken
(weathering and erosion)

glass in a bottle mould

broken bottle

pouring molten glass
(eruption)

glass swept up
(transport)

molten glass in furnace

broken bottle in recycling bin

broken bottle in dustpan

heat

melting
(melting)

glass thrown into bin
(deposition)

Having the cycle for the recycling of a glass bottle clearly identified, pupils will find that the rock cycle is very similar.

The rock cycle

Pupils have been introduced to sedimentary, igneous and metamorphic rocks, and they can be taught how these three types are all part of the recycling of rocks in the Earth's crust. However, it is not necessary at this stage to consider plate tectonic processes as the driving force behind aspects of the rock cycle.

Modern chemistry textbooks contain diagrams showing the rock cycle; Figure 6.10 shows one example. The degree of detail you use will depend on the ability range of the pupils being taught. There are many very detailed and colourful diagrams that pupils often enjoy turning into posters for display, and this may be a useful introduction for some. There are also diagrams that use boxes to represent rock types and the processes occurring.

Figure 6.10
The rock cycle.

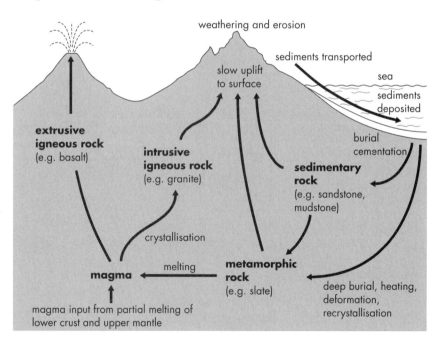

Each of these approaches to representing the rock cycle has its advantages, although it is of some interest that most examination questions set on this topic have tended to use boxes. It would therefore be advantageous for pupils to have seen the rock cycle represented in this way. You might wish to introduce pupils to both ways of representing the rock cycle.

Card games

It is possible to do an exercise with sets of cards. One set of cards has labels for rock types and another set of cards has labels for processes. Pupils can match up the rock conversions with the processes causing them. Figure 6.11 shows some examples.

Figure 6.11
Card game based on the rock cycle.

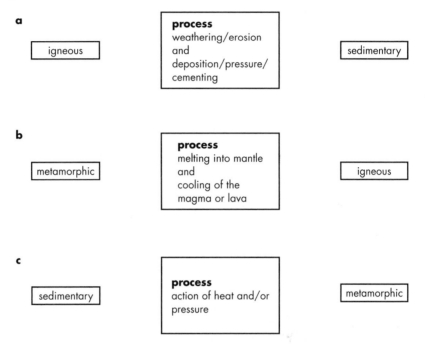

a

| igneous |

process
weathering/erosion
and
deposition/pressure/
cementing

| sedimentary |

b

| metamorphic |

process
melting into mantle
and
cooling of the
magma or lava

| igneous |

c

| sedimentary |

process
action of heat and/or
pressure

| metamorphic |

The same process cards as used in (a) could be used for the conversion of metamorphic rocks to sedimentary rocks because metamorphic rocks on the Earth's surface can also be weathered. It can also be used for sedimentary to sedimentary conversions, because uplifted sediments can be eroded to form new sediments.

In (b), it is unlikely that sedimentary rocks would melt directly into the mantle without first having gone through a metamorphic stage, considering the immense pressure that they would be under. Similarly, any intrusive igneous rock would also undergo metamorphic processes before re-melting into the mantle.

In (c), again, the same process card can be used for the conversion of igneous rock to metamorphic rock and even for metamorphic rock to metamorphic rock if the rock undergoes subsequent metamorphism.

It is important for pupils to realise that buried sediments do not automatically become part of the mantle, to re-emerge as igneous rocks: various transitions are possible.

Rock samples

The absence of any practical work in this section is partly made up for by the possibility of poster work or using the cards idea. However, now is a suitable time to examine samples of rocks from all the three main types simultaneously. There are collections of rocks available commercially. The following selection is a suggested range:

- igneous: basalt, granite
- metamorphic: slate, schist, gneiss, marble
- sedimentary: sandstone, limestone (including samples with fossils), chalk (not gypsum), mudstone, conglomerate.

Pupils should examine the rocks and be able to classify them:

- igneous: hard, made of crystals randomly arranged, no fossils, crystal size depends on the rate of cooling from molten rock
- metamorphic: hard, crystals may be in interlocking bands or layers, may cleave, fossils unlikely
- sedimentary: may be soft and crumbly (although some are harder), rounded fragments rather than crystals, may contain fossils.

♦ *Enhancement ideas*

- ♦ A study of the geological time periods, with names and notable events. A time line can be constructed on till-roll paper (a 2½ metre strip is needed by each group or each pupil) using a scale of 1 million years = 2 mm. Add on the events at appropriate places.

- ♦ Research the uses of the rocks studied in this topic.

- ♦ Examining the relative hardness of rocks and hence the use of the Moh's scale of hardness.

- ♦ Pupils may use a suitable rock key (e.g. Figure 6.12) to identify different sedimentary, metamorphic and igneous rocks.

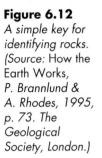

Figure 6.12
A simple key for identifying rocks. (Source: How the Earth Works, P. Brannlund & A. Rhodes, 1995, p. 73. The Geological Society, London.)

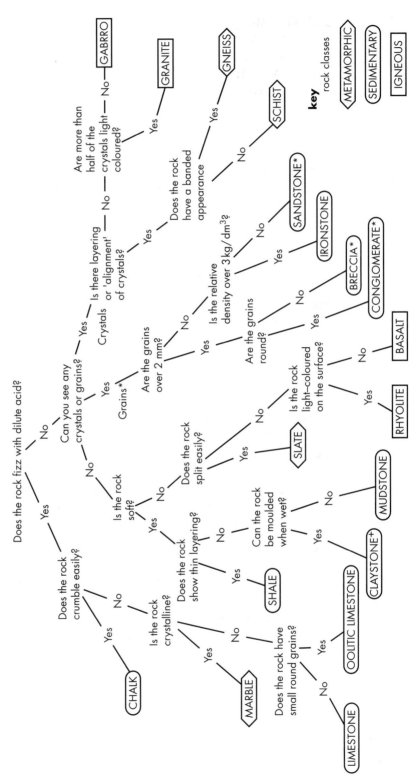

6.5 Plate tectonics

◆ *Previous knowledge and experience*

Some pupils may have a simple understanding of the internal structure of the Earth but they are unlikely to have been taught about plate tectonic processes in Science.

◆ *A teaching sequence*

The thickness of the Earth's crust was discussed on page 190, in the section on igneous rocks. Pupils now need to understand what lies beneath the crust before the concept of plate tectonics can be appreciated. Textbooks will contain suitable diagrams, but there is a certain minimum that pupils should understand. The core is made from iron and nickel; it is solid at its centre, where the pressure is greatest, but molten around its outside. Surrounding the core, but underneath the crust, is the mantle. The mantle is mostly solid, but just under the crust it can be thought of as molten. The crust is therefore resting on top of this very viscous molten rock that forms the top of the mantle.

Convection currents in the mantle

The Earth retains a hot core because of energy given off by natural radioactive processes occurring within the Earth itself. Parts of the mantle (called 'hot spots') are hotter than others, and this allows the formation of convection currents in the molten rock. There is a simple way to model a convection current.

Modelling convection currents

Materials
- eye protection
- large beaker (250 cm^3 minimum, but larger ones are much better)
- water
- potassium manganate(VII) crystals
- tweezers or spatula
- Bunsen burner

Safety
- *Potassium manganate(VII) is an oxidising agent and is harmful.*
- *Potassium manganate(VII) crystals should not be touched with fingers as they cause brown stains.*
- *Wear eye protection when handling potassium manganate(VII).*
- *Wear eye protection when using a Bunsen burner.*

Procedure
1. Fill the beaker about three-quarters full with water.
2. Gently warm the bottom of the beaker on one side.
3. Drop in two or three crystals of potassium manganate(VII) over the hot spot. Wear eye protection. (Timing is vital to set up a nice convection current. Practise before doing it in front of the class!)

What you might expect
The crystals sink, start dissolving and, if the current has formed, send trails of the dissolving potassium manganate(VII) along the path of the convection current present in the water (see Figure 6.13a).

Figure 6.13
Using potassium manganate(VII) to reveal convection currents.
a The coloured trail rises over the hot spot.
b Sometimes a single hot spot can give rise to two or more convection currents.

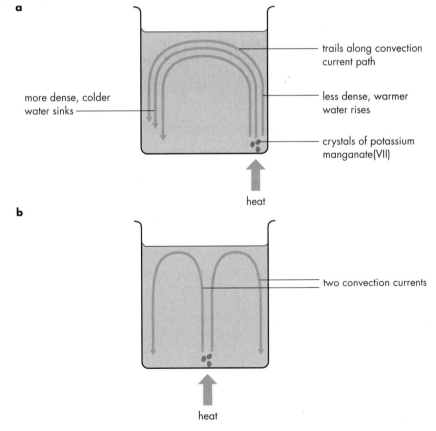

Different groups could try making hot spots at different places in their beakers. If the beaker is big enough, it may be possible to create a central hot spot and look for convection currents moving up from the centre and down both edges (see Figure 6.13b).

It is vital now to link these observations to convective processes occurring in the mantle (see Figure 6.14). If pupils can appreciate this link, the next step is to consider the effects that these currents will have on the crust which is floating on top of the mantle.

Figure 6.14
Convection currents in the Earth's mantle.

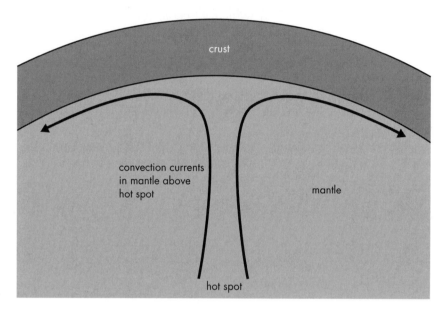

Continental crust and oceanic crust

There are two distinct sorts of crust. Continental crust has the approximate composition of granite and a density of about 2.7 g/cm^3, while oceanic crust is mostly basalt, dolerite and gabbro, with a density of about 3.2 g/cm^3. The greater density of the oceanic crust means that it floats lower down in the mantle. However, the continental crust is much thicker than the oceanic crust and so will be forced further down into the mantle by its sheer thickness and mass. These situations can be modelled.

Materials
- aquarium or fish tank, about 50 cm × 10 cm × 10 cm
- 3 wooden blocks of the same wood, each about 10 cm wide and 15 cm long but of different thicknesses (e.g. 6 cm, 4 cm, 2 cm)
- 2 wooden blocks, one of pine and one of oak, each 15 cm × 10 cm × 6 cm

Procedure
1. Put each of the different blocks of the same wood alongside one another in the tank of water. You should observe the situation shown in Figure 6.15a. This shows that the thicker the crust, the deeper it sinks into the mantle. Pupils should now be able to link the idea of the thickness of continental crust to features seen on the Earth's surface. The crust is much thicker under mountain ranges than on low-lying plains.
2. Do the same with the blocks of the two different types of wood. The denser oak sinks lower into the water than the less dense pine (Figure 6.15b).

Figure 6.15
Using wooden blocks to model how the Earth's crust floats on the mantle.
a *The thicker the block, the deeper it sinks.*
b *The denser the block, the deeper it sinks.*

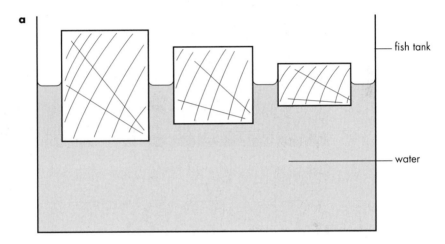

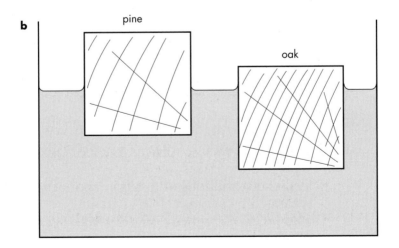

These observations can now be linked to the idea of the oceanic crust sinking further down into the mantle than the less dense continental crust (see Figure 6.16). It should now be possible for pupils to appreciate how oceanic crust and continental crust float on the mantle at different depths.

Figure 6.16
Oceanic crust is denser than continental crust, and sinks further into the mantle.

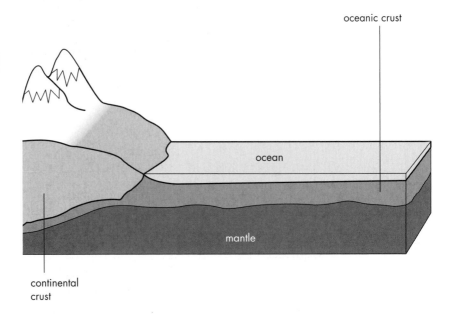

Evidence for plate tectonics

It is now possible to link the idea of convection currents in the mantle with the idea that parts of crust are floating on the mantle. These parts of crust are called 'plates' and are carried across the mantle by the convection currents below them (tectonic processes). The observable features on the surface can be explained by using this theory.

This is the reverse of how Alfred Wegener first generated his ideas about plate tectonics in the 1910s: he used the observable surface features on the Earth to come up with ideas about what might be happening underneath it. Textbooks will list the evidence:

- present-day continents fitting together like jigsaw pieces
- similarity of rock sequences in continents that are now separated by oceans
- similar distributions of fossils in continents that are now separated by oceans
- alignment of magnetic particles in the crust on either side of mid-ocean ridges (see Figure 6.17).

Figure 6.17
Magnetic 'stripes' in the ocean bed.

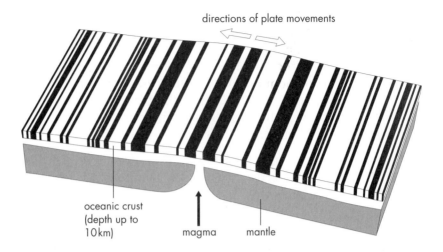

directions of plate movements

oceanic crust
(depth up to
10km)

magma

mantle

There is some scope here for a discussion about the reluctance
of scientists to accept new ideas. The previously accepted
theory for the presence of mountains (and other features) on
the Earth's surface was that the Earth was slowly cooling down
and contracting as it did so. This would produce a wrinkling of
the surface, similar to what happens when an apple withers. It
was only the discovery of the symmetrical magnetic 'stripes' on
either side of the mid-ocean ridges in the 1960s that finally
convinced scientists that Wegener's ideas were in fact correct.

Plate boundaries

Textbooks will provide world maps showing plate boundaries
and the direction of movement of these plates; Figure 6.18
shows an example.

Figure 6.18
The boundaries between the tectonic plates.

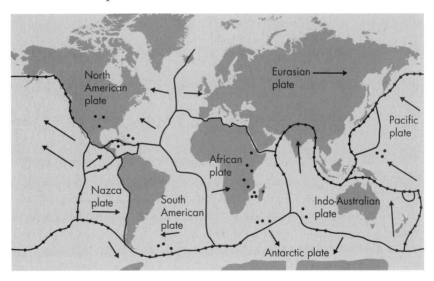

North
American
plate

Eurasian
plate

Pacific
plate

African
plate

Nazca
plate

South
American
plate

Indo-Australian
plate

Antarctic plate

Various exercises are possible, plotting volcanoes and earthquake zones on to maps showing the plate boundaries. A good way to show this is to use an overhead projector and successive overlays of maps showing plate boundaries, volcanoes and earthquake zones. A final overlay can show the relative movements of the plates. This will lead on to discussion of what happens when different sorts of plates meet.

There are four important examples that pupils should know.

- A constructive plate boundary at a mid-ocean ridge, forming new crust; oceanic crust moves away from this at equal rates on both sides.
- The collision of an oceanic plate and a continental plate; volcanoes and earthquakes occur as the denser oceanic crust is forced under the less dense continental crust, and mountain ranges form.
- The collision of two continental plates; fold mountains and earthquakes occur.
- Sideways movement of two continental plates; earthquakes occur.

Most textbooks will provide good diagrams to represent these situations.

Folding and faulting

One of the consequences of plate movement is that layers of rock in the crust become folded and/or faulted because of compression and/or tension forces.

Materials
- silver sand, 100 g
- flour, 15 g
- transparent plastic container, e.g. component drawer
- hardboard piece to fit vertically across drawer, snugly to the edges
- large spatula or spoon
- large tray to collect the 'overflow'

The quantities in the list are for a demonstration. If used as a class experiment, then these are the quantities per group. There may be a disposal problem; pupils should put used samples into a large tray; do not let the solids go into sinks. If only small amounts of flour have been used, the final sand–flour mix can be kept and used again several times.

Figure 6.19
Using a clear plastic drawer to demonstrate folding and faulting.
a *The transparent drawer with its hardboard separator.*
b *Alternate layers of sand and flour in one half of the drawer represent bands of rock.*
c *Applying gentle pressure to induce folding.*
d *Increasing the pressure may create a compression fault.*

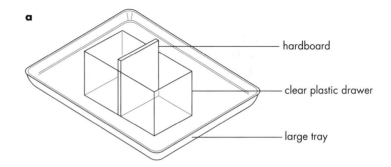

a

hardboard

clear plastic drawer

large tray

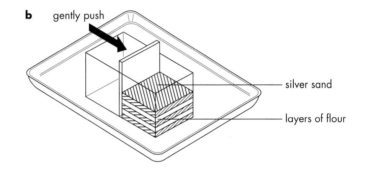

b   gently push

silver sand

layers of flour

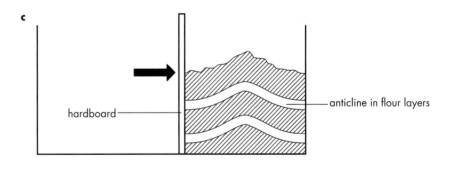

c

hardboard

anticline in flour layers

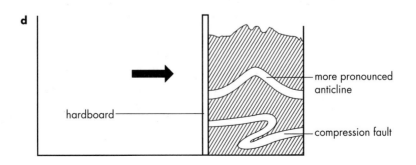

d

hardboard

more pronounced anticline

compression fault

Procedure

1. If the component drawer has a central divider, then use only half of it. If it does not have a divider, the whole drawer can be used. Put the hardboard piece vertically into the drawer (Figure 6.19a) and place the drawer on a large tray.

2. Put alternate layers of sand and flour into the drawer (Figure 6.19b). Only half-fill the drawer.

3. Carefully push the board against the sand. This represents a compressional force on the layers of rock. The layers should start to fold, producing an anticline (Figure 6.19c). Pupils can then link this to examples of folded rock layers in the Earth's crust. A diagram in a textbook or photographs would be useful.

4. It is sometimes possible to create a compression fault. The fault angle may vary and may be almost horizontal. Sometimes a fold and fault may be found in the same experiment (Figure 6.19d).

Pupils can sketch what they see when different amounts of compression are applied to the 'rocks' using the hardboard.

It is important to point out that although these are not real layers of rock, the layers of sand and flour do behave in a similar way. Notice how the sideways pressure lifts the 'rock layers' up.

Pupils should now appreciate how sedimentary rocks that were formed below sea level can be found millions of years later at the top of mountains: compression of layers causes uplift. The exact geometry of folds and faults can be used by geologists to tell the direction and strength of the forces in the crust which produced the observed faults and folds.

◆ *Enhancement ideas*

◆ Compression faults are often almost horizontal. Where plates are moving apart (forming in the first instance a rift valley), the tensional stresses produce steeper faults. Pupils might be asked how they could try to model this using apparatus similar to that used above. Hopefully, they will suggest putting layers between two pieces of hardboard and then slowly moving the hardboard pieces apart. This may not always give convincing results, however.

Figure 6.20
Using Plasticine to model folding, faulting, erosion and deposition.
a *The Plasticine layers.*
b *Applying gentle pressure to induce folding.*
c *Increasing the pressure may create a fault.*
d *Modelling erosion by cutting off the top.*
e *Modelling subsequent deposition to produce an unconformity.*

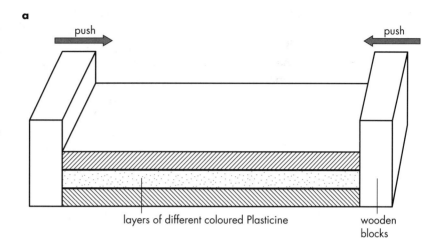

a

push   push

layers of different coloured Plasticine   wooden blocks

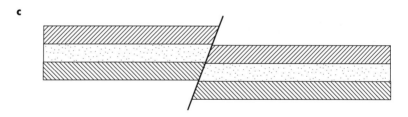

b

push   push

anticline forms

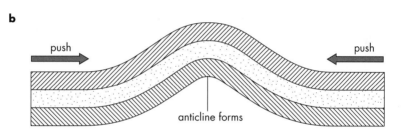

c

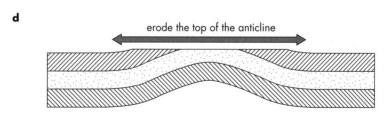

d

erode the top of the anticline

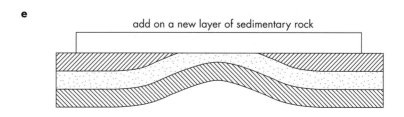

e

add on a new layer of sedimentary rock

◆ Folding and faulting can also be modelled using layers of different coloured Plasticine (see Figure 6.20a). Compress the Plasticine with wooden blocks to produce an anticline (Figure 6.20b). If the Plasticine is not very pliable and is rather rigid, it may be possible to simulate a fault instead (Figure 6.20c). If this fails, try cutting the Plasticine and moving one layer down. Other possibilities then include 'eroding' the top surfaces of these folded/faulted structures (by cutting the top off with a knife) (Figure 6.20d), then putting new layers of Plasticine (new 'sediments') on top, creating an unconformity (Figure 6.20e).

◆ Earthquakes are common along plate boundaries and it is possible to make simple seismometers, model the effect of earthquakes on buildings that are built on soft ground and look at the different sorts of shock waves. Teachers should consult the excellent book *Earth Science: Activities and Demonstrations* by Mike Tuke.

◆ *Other resources*

 The *Encarta World Atlas* has a good section on the structure of the Earth, including text, pictures and video clips. Particularly useful is a world map showing Earth plates and sites of volcanic activity. Pupils can, for example, look at Iceland and zoom in to see the plate boundary and volcanic sites close by.

Many schools and colleges do not have a large range of rock samples for pupils to look at. Rocks and equipment can be obtained from the following suppliers:

Geosupplies, 16 Station Road, Chapeltown, Sheffield S30 4XH. Equipment and geological books.

Michael Jay Publications (MJP) – Geopacks, PO Box 23, St Just, Penzance, Cornwall TR19 7JS. Equipment, transparencies, computer software, visual aids.

Offa Rocks, Lower Hengoed, Oswestry, Shropshire SY10 7AB. Rocks, mineral samples and fossils.

Richard Tayler, Byways, 20 Burstead Close, Cobham, Surrey KT11 2NL. Rock and mineral samples.

Griffin and George, Bishops Meadow Road, Loughborough, Leicestershire LE11 0RG. General science equipment, including rock sets, geology tools, kits, etc.

Philip Harris, Lynn Lane, Shenstone, Lichfield, Staffordshire WS24 0EE. General science equipment.

Alternatively, samples of rock can be obtained free from stonemasons' yards. Local quarries and cliffs may be other sources. Contact a local school, college or university which teaches Geology A level; they may have surplus samples of rocks.

Web sites
♦ Information about plate tectonics can be found on:
   www.hcrhs.hunterdon.k12.nj.us/science/ptech.html
♦ Information on recent earthquakes and volcanoes can be found on:
   www.artbell.com/earthquakes.html

Further reading
The Earth Science Teachers Association, 1996. *Investigating the Science of the Earth*. ISBN 1 8732 6612 X.
   SoE1 Changes to the atmosphere
   SoE2 Geological changes – Earth's structure and plate tectonics
   SoE3 Geological changes – rock formation and deformation
Brannlund, Peter & Rhodes, Alan, 1995. *How the Earth works*. The Geological Society. ISBN 1 8977 9951 9.
Tuke, Mike, 1991. *Earth Science: Activities and Demonstrations*. John Murray, London. ISBN 0 7195 4951 5.
Webster, D., 1987. *Understanding Geology*. Oliver & Boyd. ISBN 0 0500 3664 5.
van Rose, Susanna and Mercer, Ian F., 1999. *Volcanoes*, revised 2nd edition. The Natural History Museum, London. ISBN 0 565 09138 7.

♦ *Useful addresses*

Earth Science Teachers' Association (ESTA), Burlington House, Piccadilly, London W1V 0BN.
ESTA Promotions, 4 Wyvern Gardens, Dore, Sheffield S17 3PR.

7 *The Periodic table*

Raymond Oliver

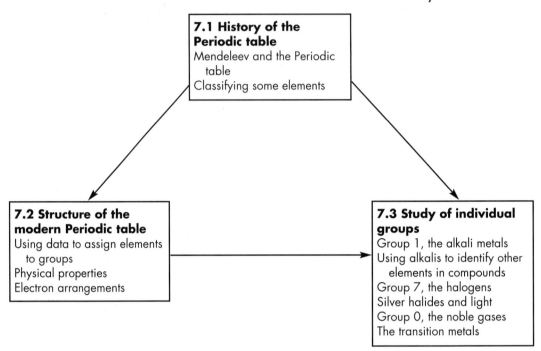

7.1 History of the Periodic table
Mendeleev and the Periodic table
Classifying some elements

7.2 Structure of the modern Periodic table
Using data to assign elements to groups
Physical properties
Electron arrangements

7.3 Study of individual groups
Group 1, the alkali metals
Using alkalis to identify other elements in compounds
Group 7, the halogens
Silver halides and light
Group 0, the noble gases
The transition metals

◆ *Choosing a route*

Pupils will be familiar with the Periodic table from seeing it on the wall of the laboratory. As they make progress through Chemistry they should find that much of their knowledge about the chemistry of the elements can be related to the position of the elements in the Periodic table.

In the suggested route, the history of the Periodic table and its structure are used as precursors to a detailed study of selected groups:

The route takes us from a general overview of the Periodic table classification to a detailed study of four groups of elements. In studying the Periodic table, pupils must be clear about the distinctions between elements, compounds and mixtures (see Chapter 1). Pupils who are unclear about these basic differences will be confused about what can and cannot be included in the table.

7.1 History of the Periodic table

♦ *Previous knowledge and experience*

Pupils will have experience of classifying everyday materials on the basis of their properties. They will not be so familiar with the distinctions between elements, compounds and mixtures (see Chapter 1).

♦ *A teaching sequence*

Mendeleev and the Periodic table

Pupils encountering the Periodic table for the first time are both intrigued and puzzled. You can encourage them to comment on its unusual layout – a series of boxes with a gap in the middle. Pupils will realise that it represents some kind of classification but not a simple one. The historical development of the Periodic table teaches pupils some important lessons about the nature of science.

By the end of the 18th century about 30 elements were known. As further elements were isolated and identified, it became clear that some were closely related in their chemical and physical properties. Chemists attempted to look for patterns in the elements. They had accurate atomic masses for the elements. Since most elements had atomic masses close to whole numbers, Prout advanced a hypothesis that all elements were formed by the coalescence of hydrogen atoms. Pupils could look at a list of elements and their relative atomic masses and you could discuss which elements do not fit this hypothesis. With more able pupils you could remind them of different isotopes of the same element: this is the explanation for why not all relative atomic masses are close to whole numbers.

In 1829 Johan Döbereiner noticed that some groups of three elements could be found with similar properties and with the atomic mass of the middle one an average of the other two, e.g. lithium, sodium and potassium; calcium, strontium and barium; and chlorine, bromine and iodine. He called a group of three such elements a 'triad'. Let the pupils check the atomic masses of these elements and look for other triads.

A school Chemistry textbook of 1841 (*Manual of Chemistry* by W. T. Brande) listed 55 elements but with the following caveat: 'it may be presumed that some are in reality compounds and will be proved to be so as science advances'. Brande reminds his readers that it is 'experiment and not hypothesis that is resorted to as the ultimate test'.

In 1869 John Newlands was the first to notice that if the elements were written down in ascending order of atomic mass, similar properties recurred in every eighth element, like the octaves of a musical scale. This worked well for the lighter elements but broke down with heavier elements.

By 1869 the Russian chemist Dmitri Mendeleev had published a Periodic table of all known elements. He realised that further elements were likely to be discovered and left gaps in the table where he considered this to be appropriate. The key features of Mendeleev's classification were as follows:

- similar elements were grouped together
- elements were displayed in the order of their atomic masses
- where anomalies would otherwise occur, Mendeleev exchanged the positions of the elements. For example, iodine and tellurium are in the 'wrong' places from what would be expected from their masses
- gaps were left where Mendeleev suspected that an element was yet to be discovered, for example, the semi-metal now called germanium.

Figure 7.1 shows a copy of Mendeleev's original table of 1869. Pupils could count the elements in his table. Some elements he showed with question marks; what names do these elements now have? In what ways is his table similar to our modern table and in what ways is it different?

The table in Figure 7.2 (see page 220) was printed in a chemistry book by Sir Henry E. Roscoe, Professor of Chemistry at Victoria University in Manchester and leading author of chemistry books in the 19th century. He wrote this book in 1882. Pupils could compare this table with Mendeleev's original table and with our modern Periodic table.

Figure 7.1
Dmitri Mendeleev's Periodic table of the elements, 1869.

но въ ней, мнѣ кажется, уже ясно выражается примѣнимость выставляемаго мною начала ко всей совокупности элементовъ, пай
которыхъ извѣстенъ съ достовѣрностію. На этотъ разъ я и желалъ
преимущественно найдти общую систему элементовъ. Вотъ этотъ
опытъ:

```
                              Ti=50     Zr=90      ?=180.
                              V=51      Nb=94     Ta=182.
                              Cr=52     Mo=96      W=186.
                              Mn=55     Rh=104,4  Pt=197,4
                              Fe=56     Ru=104,4  Ir=198.
                          Ni=Co=59      Pl=106,6  Os=199.
   H=1                         Cu=63,4   Ag=108   Hg=200.
         Be=9,4   Mg=24        Zn=65,2   Cd=112
         B=11     Al=27,4       ?=68     Ur=116   Au=197?
         C=12     Si=28         ?=70     Sn=118
         N=14     P=31        As=75      Sb=122   Bi=210
         O=16     S=32        Se=79,4    Te=128?
         F=19     Cl=35,5     Br=80      I=127
   Li=7  Na=23    K=39        Rb=85,4    Cs=133   Tl=204
                  Ca=40       Sr=87,6    Ba=137   Pb=207.
                   ?=45       Ce=92
                  ?Er=56      La=94
                  ?Yt=60      Di=95
                  ?In=75,6    Th=118?
```

а потому приходится въ разныхъ рядахъ имѣть различное измѣненіе разностей,
чего нѣтъ въ главныхъ числахъ предлагаемой таблицы. Или же придется предполагать при составленіи системы очень много недостающихъ членовъ. То и
другое мало выгодно. Мнѣ кажется притомъ, наиболѣе естественнымъ составить
кубическую систему (предлагаемая есть плоскостная), но и попытки для ея образованія не повели къ надлежащимъ результатамъ. Слѣдующія двѣ попытки могутъ показать то разнообразіе сопоставленій, какое возможно при допущеніи основнаго
начала, высказаннаго въ этой статьѣ.

```
Li   Na    K    Cu    Rb    Ag    Cs    —    Tl
 7   23   39   63,4  85,4  108   133        204
Be   Mg   Ca   Zn    Sr    Cd    Ba    —    Pb
 B   Al    —    —     —    Ur     —    —    Bi?
 C   Si   Ti    —    Zr    Sn     —    —     —
 N    P    V   As    Nb    Sb     —   Ta     —
 O    S    —   Se     —    Te     —    W     —
 F   Cl    —   Br     —     J     —    —     —
19  35,5  58   80    190   127   160  190  220.
```

Figure 7.2
*Part of Henry
Roscoe's Periodic
table, 1882.*

Groups	I.	II.	III.	IV.	V.	VI.	VII.	VIII.
Series.	—	—	—	RH_4	RH_3	RH_2	RH	(R_2H) Hydrogen
	R_2O	R_2O_2 or RO	R_2O_3	R_2O_4 or RO_2	R_2O_5	R_2O_6 or RO_3	R_2O_7	Compounds. R_2O_8 or RO_4} Higher Oxygen Compounds
1 2	1 H Li 7	Be 9	B 11	C 12	N 14	O 16	F 19	
3 4	23 Na K 39	24 Mg Ca 40	27 Al Sc 44	28 Si Ti 48	31 P V 51	32 S Cr 52	35.5 Cl Mn 55	Fe 56. Co 59. Ni 59.

The modern form of the Periodic table is very similar to
Mendeleev's prototype. It includes both the naturally
occurring elements and those that have been made artificially
by nuclear reactions, such as promethium in 1945. The
discovery of the electron and an understanding of the rules
governing the arrangement of electrons in atoms, explains the
modern form of the Periodic table. From left to right across
each horizontal row (or 'period'), one energy level is
progressively filled up with electrons. In the following period,
the next highest energy level is filled with electrons. Similar
electronic arrangements in different atoms account for
similarities in their chemical and physical properties. For
example in Group 1, the alkali metals, the electronic
configurations are as shown in Table 7.1. All of the elements
in Group 1 have one electron in their outer shell (or highest
energy level).

Table 7.1 *Electronic configurations in Group 1, the alkali metals.*

Element	Proton number	Electron configuration
Lithium	3	2,1
Sodium	11	2,8,1
Potassium	19	2,8,8,1

Classifying some elements

Materials
- battery and holder
- wires
- bulb
- mercury thermometer
- clear light bulb
- wood splints
- aluminium foil
- stoppered tube of oxygen
- sealed ampoule of bromine
- sealed tube containing iodine crystals
- copper foil
- roll sulphur

Safety
- *Sulphur is a fine powder which irritates the eyes.*
- *Oxygen is very oxidising.*
- *Mercury is toxic. Electrical conductivity test must be carried out in a fume cupboard.*
- *Bromine has an irritant vapour. It causes severe burns and it is very toxic by inhalation.*
- *Iodine is harmful by inhalation and skin contact and causes burns over time.*
- *Iodine vapour crystallises very painfully on eyeballs.*
- *Wear eye protection when handling hazardous chemicals.*

Procedure
1. Test the electrical conductivity of aluminium, copper and sulphur by using the battery, wires and bulb (Figure 7.3, overleaf). As a teacher demonstration, test the electrical conductivity of mercury. This *must* be carried out in a fume cupboard.
2. Test the oxygen by placing a glowing splint into the tube of gas.
3. Examine the enclosed elements – mercury, argon in the clear light bulb, liquid bromine and the iodine crystals.
4. Classify the sample elements as solid, liquid or gas, and as conductor or insulator. What is special about metals?

Figure 7.3
*Testing electrical
conductivity: if the
sample conducts
electricity, the
lamp will light.*

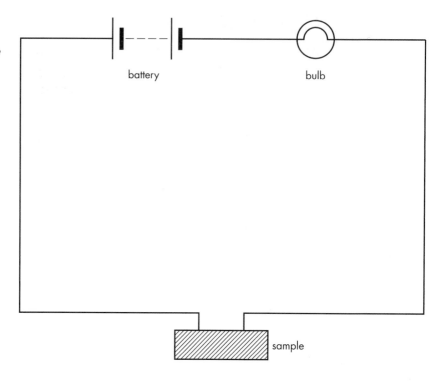

battery   bulb

sample

What you might expect
Metals will conduct electricity and the lamp will light. Mercury will also
give this result even though it is a liquid.

The non-metals are insulators; they are argon, oxygen (the splint
re-lights, showing the invisible oxygen really is present), bromine, iodine
and sulphur.

Dividing the elements into two categories of metals and non-metals is
not as simple as it seems. Metals are usually shiny, conductors, flexible,
solid and dense. There are exceptions, however, such as sodium, which is
soft and floats on water, and mercury, which is a liquid. In the Periodic
table metals are found on the left and non-metals on the right. Some
elements have properties that are intermediate – the semi-metals such as
silicon and germanium in Group 4.

◆ *Enhancement ideas*

- ◆ Test the electrical conductivity of elements we use in everyday life, using the same battery, wires and lamp. For example, try gold jewellery, a nickel spatula, a silver spoon, carbon in an HB pencil (graphite, a non-metal which conducts), an iron nail, a tin-plated food can, a galvanised bucket (steel coated in zinc). Locate each element in the Periodic table.
- ◆ Carry out a data search on a selection of elements to look at the variation in other properties such as their melting point, density or thermal conductivity (see Figure 7.4).

- ◆ Plot graphs of physical properties such as melting point, boiling point or density against atomic number (proton number). These graphs could be drawn using a graph drawing package. Look for periodic graphs which are a series of peaks and troughs (see Figure 7.4). Which elements are in the peaks and which are in the troughs? Are these elements in the same group of the Periodic table? Examine graphs of other properties against atomic number.

Figure 7.4
Physical properties of the elements hydrogen to argon.
a Melting point.
b Boiling point.

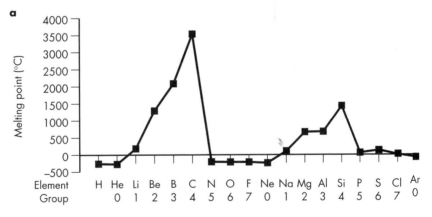

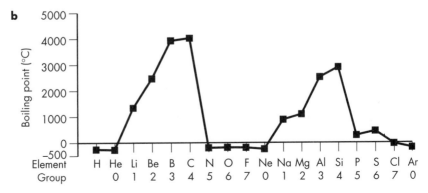

7.2 Structure of the modern Periodic table

◆ *Previous knowledge and experience*

Older pupils who are familiar with the details of atomic structure (see Chapter 2) will see that the Periodic table arranges elements by atomic *number* (proton number). This contrasts with Mendeleev's sequence by atomic *mass*, although the two sequences are very similar.

◆ *A teaching sequence*

Pupils should understand that a periodic property is one that recurs at regular intervals. In the 19th century, scientists noticed that elements with similar properties occurred after every eighth element, for example sodium (number 11) and potassium (number 19). We now know this to be a consequence of the way in which electrons are arranged in atoms.

Pupils need to know that:

- similar elements are placed in the same vertical column, called a 'group'
- horizontal rows are called 'periods', with metals on the left of the table
- groups and periods are allocated numbers
- some groups have special names, such as the alkali metals of Group 1.

Figure 7.5 shows the modern Periodic table.

Figure 7.5
The modern Periodic table.

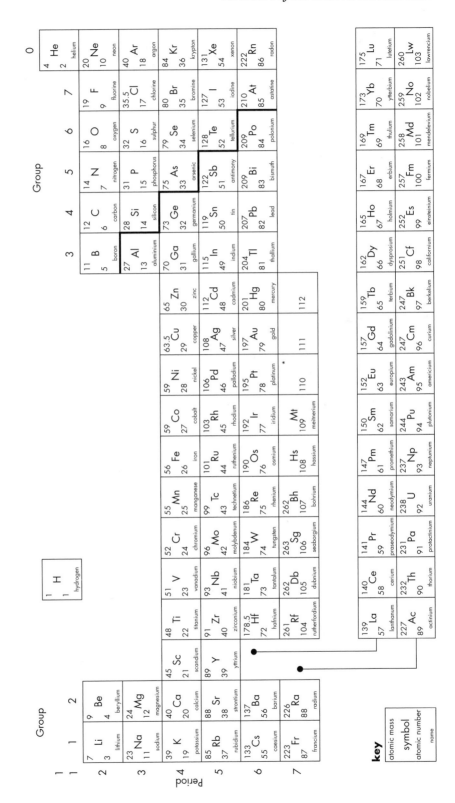

Using data to assign elements to groups

Materials

Prepare element data cards, the size of playing cards, for Groups 1 (the alkali metals), 7 (the halogens) and 0 (the noble gases), with the following information: name, symbol, state (solid, liquid or gas), density (floats or sinks in water), how it is stored (e.g. under oil), how it reacts with water or steam, ionic charge, formula of an oxide and a chloride (see Figure 7.6).

Provide samples of the elements, where possible, or photographs.

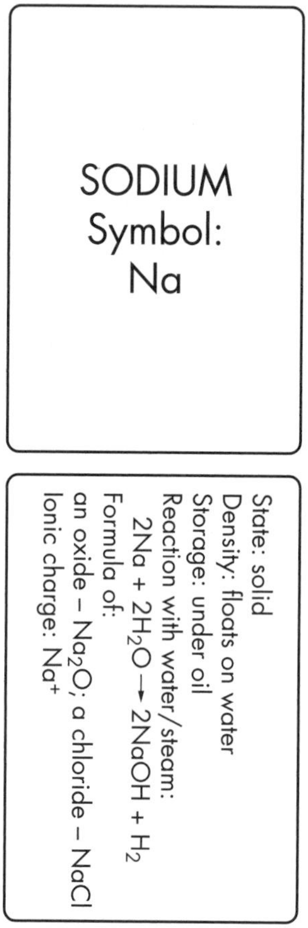

Figure 7.6
Data card for sodium.

SODIUM
Symbol:
Na

State: solid
Density: floats on water
Storage: under oil
Reaction with water/steam:
$2Na + 2H_2O \rightarrow 2NaOH + H_2$
Formula of:
an oxide – Na_2O; a chloride – $NaCl$
Ionic charge: Na^+

Procedure

1. Shuffle the cards.
2. Working in groups, examine the element cards and look for similarities between elements.
3. Place similar elements in the same groups.
4. Compare the outcome with a copy of the Periodic table.
5. Comment on any differences, for example that the same elements are present but there is no evidence for the order within each group.

Physical properties

The most obvious trend in the Periodic table is that metals
appear on the left and non-metals on the right. Pupils should
be clear that metals are elements which show a combination of
properties, which typically include good electrical and thermal
conductivity, a characteristic surface shine described as metallic
'lustre', malleability (can form thin foils), ductility (can form
wires and tubes) and high density. Non-metals do not usually
show such properties, but the distinction between metals and
non-metals is an ambiguous one. For example, the metal
mercury is a liquid, the metal sodium floats on water, and the
non-metal graphite (carbon) conducts electricity.

There is a continuum in the properties of elements on
moving from the left to the right of the Periodic table. Within
each period there is a change from metallic to non-metallic
properties as you move to the right. The usual dividing line
runs diagonally from the element boron in Group 3 down to
astatine at the base of Group 7, the halogens. Elements
adjacent to this dividing line have been called 'semi-metals' or
'metalloids'; examples include silicon, germanium and arsenic.

Pupils could arrange a list of elements as metals or non-
metals using the position of the elements in the Periodic table.
The most obviously non-metallic elements are found in the top
right-hand corner of the Periodic table. Elements here include
the noble gases of Group 0, the halogens of Group 7 and other
familiar non-metals such as oxygen, sulphur and nitrogen.

Melting points give a useful indication of the type of
element. The melting points rise across a period from the
metals to the metalloids and then fall again on reaching the
non-metals:

- in Group 1, the alkali metals, the melting point and boiling
  point both become lower as you go down the group
- in Group 7, the halogens, the melting point and the boiling
  point both get higher (gas, gas, liquid, solid) as you go
  down the group.

There are also trends in the density of elements. The density
of elements, measured in g/cm^3, increases from the left of a
period, reaching a maximum in Group 4, for example with
carbon. The maximum value corresponds to small atomic
radius coupled with strong giant covalent or metallic bonding.

Electron arrangements

Within a particular group, the outer shell of the different elements' atoms contains the same number of electrons. For example, in the alkali metals of Group 1, all of the elements have one electron in their highest occupied energy level (outer shell). In Group 7, the halogens, all the elements have seven electrons in their outer shell. The trend across the Periodic table within a single period is that each successive element adds one more electron to the highest occupied energy level. When this level is full, we reach Group 0, the noble gases. Elements in the same group are similar because they have similar electron arrangements. The higher the energy level, the more easily the outer electrons can be lost (for metals), and the less easily extra electrons can be gained (for non-metals).

The noble gases of Group 0 are both unreactive and monatomic (the molecule is a single atom) because the top occupied energy level is already full. Noble gas atoms do not gain, lose or share electrons under ordinary circumstances.

Group 1 elements have one electron more than a full shell and tend to lose one electron to give positive ions. Group 7 elements (halogens) need one more electron to complete their outer shell. Group 7 elements tend to gain one electron to form negative ions or share an electron to form a covalent bond.

Pupils should work out the electron arrangements of elements in period 2 or 3 and in Group 1 or 2 (see Figure 7.7). Although the elements within a single group are similar, they are not the same. The elements show trends in their properties. For example, they may become more reactive as you descend or ascend a group. Two groups of elements are especially suitable for demonstrating trends in properties; Group 1, the alkali metals, and Group 7, the halogens. Pupils who establish that elements increase in reactivity down Group 1 often assume that a similar pattern exists in all other groups. This leads to mistakes in predicting the relative reactivities of the elements in Group 7.

Figure 7.7
Electronic structures of the 20 lightest elements.

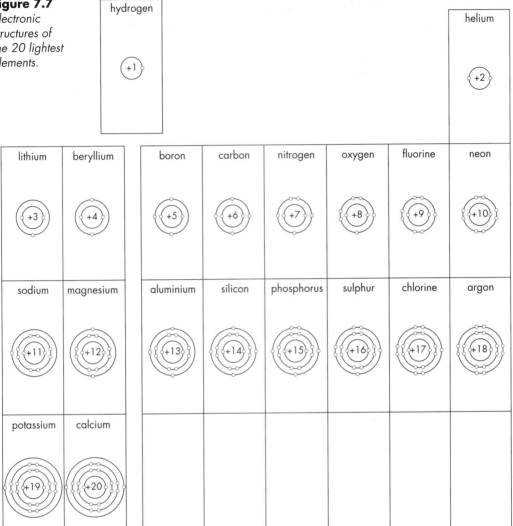

♦ *Enhancement ideas*

♦ There are many different forms of the Periodic table. Find other examples of Periodic tables. What are the advantages and disadvantages of these different Periodic tables?

♦ Draw up a Periodic table with small diagrams of the electronic structures of the atoms of each of the elements.

7.3 Study of individual groups

Group 1, the alkali metals

Trends in the properties of the alkali metals

Remind pupils of the differing reactivities of the alkali metals as demonstrated in Chapter 4, page 108. Table 7.2 summarises the changes to the metals and compares their activity.

Table 7.2 *Reactions of the alkali metals with water, with phenolphthalein indicator present.*

Metal	Changes	Relative activity
Lithium	Floats, fizzes, releases hydrogen, moves about, indicator goes pink, alkali formed	Third place
Sodium	As before, but metal melts to a bead	Second place
Potassium	As sodium, but hydrogen gas burns with lilac flame, can explode on water surface	First place

In every case we have

metal + water → hydrogen gas + metal hydroxide (an alkali)

The alkali metals all produce solutions of alkalis and release hydrogen gas. The trend is that the reactivity increases as you go down Group 1: caesium would explode on contact with water. The equations all have the same form, for example

$$2Na + 2H_2O \rightarrow 2NaOH + H_2$$

Flame colours of alkali metal compounds

The lilac colour of the flame with potassium leads on to a consideration of alkali metal flame colours. The compounds of the alkali metals impart different but characteristic colours to flames. These find applications in fireworks, flares and in sodium vapour street lamps (orange colour).

Materials
- eye protection
- 6 nichrome flame wires, 5 cm long, each fused into the end of a glass rod handle
- 6 watch glasses
- Bunsen burner
- cobalt blue glass
- solid samples of the chlorides and bromides of lithium, sodium, potassium, about 0.5 g of each

Equipment note
Nichrome wires become corroded after repeated use and progressively more difficult to clean. You can clean them by dipping them into a small amount of concentrated hydrochloric acid and then heating again. This is unpleasant, since the acid releases hydrogen chloride fumes. Platinum wires are better but expensive.

Safety
- *Lithium chloride and lithium bromide are irritants.*
- *Good ventilation is needed as the compounds are vaporised in the flame.*
- *Wear eye protection when using a Bunsen burner.*

Procedure
1. Using a hot flame, locate the hottest part of the gas flame by lowering the wire into it. (Look for the brightest red glow.)
2. Dip the end of the hot wire into a sample solid on a watch glass.
3. Place the wire and sample into the hottest part of the flame.
4. Note the first flash of colour.
5. Repeat with the other samples, using a separate wire for each type to remove the need to clean wires and to avoid contaminating samples.

What you might expect

Table 7.3 *Flame colours of alkali metal compounds.*

Compound	Flame colour seen
Lithium compounds	Red
Sodium compounds	Bright yellow
Potassium compounds	Lilac (crimson if viewed through blue glass)

The colour comes from the metal ion, not from the anion (chloride or bromide). Potassium compounds are often contaminated by traces of sodium compounds, and the more intense sodium colour can mask the potassium flame colour. Cobalt blue glass cuts out the yellow colour from the sodium, allowing the potassium colour to show as crimson.

Using alkalis to identify other elements in compounds

The alkalis that are produced when Group 1 elements react with water find uses in analysis, because many solutions of metal compounds react with alkalis to give solid products (precipitates) which have distinctive colours.

Materials
- eye protection
- 6 test-tubes, 100 × 16 mm
- test-tube rack
- stirring rod
- dropping pipettes for each solution used
- sodium hydroxide solution (0.1 mol/dm³)
- solutions containing the following cations: copper(II), iron(II), iron(III), nickel(II), magnesium, aluminium(III), all 0.1 mol/dm³

Safety
- *0.1 mol/dm³ sodium hydroxide is an irritant.*
- *Pupils often think that alkalis such as sodium hydroxide are less dangerous than acids, but this is not the case.*
- *Eye protection must be worn at all times and the concentrations stated should not be exceeded.*

Procedure
1. Pour about 2 cm depth of one metal solution into a tube.
2. Add about ten drops of sodium hydroxide and note any changes.
3. Add about 3 cm depth of sodium hydroxide solution to the same tube and stir with care.
4. Look for colour changes, precipitates and whether a precipitate re-dissolves in excess alkali.

What you might expect

Table 7.4 *Some precipitates formed with sodium hydroxide solution.*

Metal cation	With sodium hydroxide	With excess sodium hydroxide
Copper(II)	Blue	Insoluble
Iron(II)	Green, goes brown in air	Insoluble
Iron(III)	Brown	Insoluble
Nickel(II)	Green	Insoluble
Magnesium	White/colourless	Insoluble
Aluminium	White/colourless	Soluble

The second addition of sodium hydroxide is to check the solubility of the precipitate in excess alkali; this allows pupils to distinguish between magnesium and aluminium.

The metal cations react with hydroxide ions to form metal hydroxides. The distinctive colours and behaviour with excess alkali allow pupils to identify unlabelled ions in solution. For example:

$$FeSO_4 + 2NaOH \rightarrow Fe(OH)_2 + Na_2SO_4 \quad \text{or}$$
$$Fe^{2+} + 2OH^- \rightarrow Fe(OH)_2$$

All of the cations which give coloured precipitates are transition metals (see below). The cations of Group 1 (the alkali metals) and Group 2 are colourless, as is aluminium in Group 3.

Group 7, the halogens

The halogen elements of Group 7 are all non-metals. They form negative ions (anions) and react with metals to form salts. For example, sodium burns in chlorine to give sodium chloride, common salt. The trend of reactivity in Group 7 is exactly the opposite to that in Group 1, the alkali metals. The most reactive halogen is the first member of the group, fluorine, and the halogens become less reactive as you descend Group 7 (i.e. with increasing atomic number).

The halogen elements are all diatomic (two atoms make a molecule) and can displace less reactive halogens from their compounds. For example:

chlorine + sodium bromide → bromine + sodium chloride

Chlorine has displaced bromine because chlorine is more reactive than bromine.

Displacement reactions of the halogens

This experiment will provide pupils with enough information to establish the trend of reactivity within Group 7. The presence of a halogen can be established by the colour of its solution in a silicon-based solvent such as volasil. The colour key is shown in Table 7.5.

Table 7.5 *Colours of the halogens in volasil.*

Halogen	Colour in volasil
Chlorine	Pale green, almost colourless
Bromine	Red–brown
Iodine	Purple

The volasil takes no part in the reaction but simply serves to concentrate the halogen colour, allowing easier identification.

Materials
- eye protection
- 6 test-tubes, 100 × 16 mm with rubber stoppers
- test-tube rack
- 6 dropping pipettes
- chlorine water
- bromine water
- solution of iodine in potassium iodide
- solutions of potassium chloride, potassium bromide and potassium iodide, all 0.1 mol/dm³
- volasil (available from BDH Chemicals, Poole, Dorset)

Safety
- *Chlorine water gives off chlorine, which is a toxic gas.*
- *1% bromine water (0.06 mol/dm³) is toxic and corrosive; 0.1% bromine water (0.006 mol/dm³) is harmful and an irritant. Use a very dilute solution to avoid problems.*
- *A solution of iodine in potassium iodide stronger than 1 mol/dm³ is harmful.*
- *The halogens must be handled carefully, preferably in a fume cupboard.*
- *Eye protection must be worn when using halogens.*

Procedure
1. Add six drops of chlorine water to two tubes containing 2 cm depths of potassium bromide and potassium iodide solution, respectively.
2. Add 0.5 cm depth of volasil to each tube.
3. Stopper the tubes and shake. Leave to settle in a test-tube rack.
4. Look at the colour of the volasil layer and identify the halogen present, using the colour key.
5. Repeat for bromine water with potassium chloride and potassium iodide.
6. Repeat for iodine solution with potassium chloride and potassium bromide.

What you might expect

Table 7.6 *Halogen displacement reactions.*

Mixture	Colour of the volasil layer	Halogen present
Chlorine + potassium bromide	Red–brown	Bromine
Chlorine + potassium iodide	Purple	Iodine
Bromine + potassium chloride	Red–brown	Bromine
Bromine + potassium iodide	Purple	Iodine
Iodine + potassium chloride	Purple	Iodine
Iodine + potassium bromide	Purple	Iodine

The halogens can displace other halogens that are further down Group 7 (see Table 7.5). The results show that chlorine can displace both bromine and iodine from their salts. Bromine can displace iodine but iodine cannot displace the other halogens.

The equations for these displacement reactions take the form

chlorine + potassium bromide → bromine + potassium chloride
$$Cl_2 \ + \quad 2KBr \qquad\quad \rightarrow \quad Br_2 \ + \quad 2KCl$$

The other equations have the same form as this one.

Where displacement has failed, the colour of the original halogen remains in the volasil, for example with bromine and potassium chloride. Pupils can now draw up an activity series for the halogens, starting with the most reactive halogen. Point out that we do not use fluorine in the laboratory as it is so hazardous.

Silver halides and light

Silver halides are 'photosensitive': light causes them to decompose to give grains of silver metal:

$$2AgCl(s) \rightarrow 2Ag(s) + Cl_2(g)$$

This reaction forms the basis of black and white photography.

This experiment allows pupils to prepare silver halides and to compare the effects on them of sunlight and a magnesium flame.

Materials
- eye protection
- 2 test-tubes, 100 × 19 mm
- test-tube rack
- dropping pipettes
- tongs
- cardboard, approximately 10 × 10 cm
- filter funnel and paper
- magnesium ribbon, 5 cm
- solutions of silver nitrate, potassium chloride and potassium bromide, all 0.1 mol/dm^3

Safety
- *Magnesium is highly flammable.*
- *0.1 mol/dm^3 silver nitrate is relatively safe, but skin contact should still be avoided.*
- *Burning magnesium produces a very bright flame. This should not be looked at directly.*
- *Wear eye protection when burning magnesium.*

<u>Procedure</u>

1. Mix separate $2\,cm^3$ samples of each potassium halide with an equal volume of silver nitrate solution.
2. Place one sample in direct sunlight and leave for 30 minutes.
3. Filter the other sample.
4. Open out the filter paper with the silver halide on top and spread it on the bench. Cover one half with the cardboard to exclude light.

Figure 7.8
Exposing a silver halide to the bright light from burning magnesium.

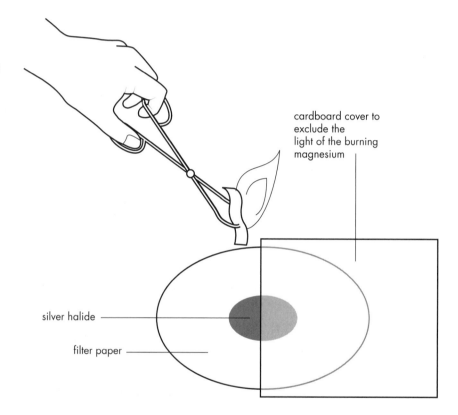

cardboard cover to exclude the light of the burning magnesium

silver halide

filter paper

5. Ignite the magnesium and hold it near to the exposed half of the residue (see Figure 7.8).
6. Remove the card and examine for any changes.

<u>What you might expect</u>
The silver halide left in sunlight turns purple and then grey–black as the sunlight reduces the halide to silver metal. The intense light of the burning magnesium causes the same change more rapidly: the exposed half turns grey, the covered half remains unchanged.

The distinctive colours of the silver halides (see Table 7.7) can also be used in analysis to identify an unknown halide ion. The aqueous halide is reacted with silver nitrate solution and the colour of the silver halide noted.

Table 7.7 *Colours of the silver halides.*

Silver halide	Colour
Silver chloride	White
Silver bromide	Cream
Silver iodide	Yellow

Group 0, the noble gases

The outer electron energy level is full in each of the noble gases of Group 0. This accounts for their stability and lack of chemical reactivity. The term 'inert gases' used to be used for Group 0, but this is inappropriate since several compounds have now been prepared. The most common compounds are the fluorides of xenon.

Helium is the second most abundant element in the universe (the most common being hydrogen). Helium occurs both in the air and in certain natural gas deposits.

With the exception of radioactive radon, all of the other noble gases occur exclusively in the air, argon representing almost 1% of air by volume.

Radon is produced by the decay of other radioisotopes, for example uranium minerals in granite rock. In the UK, the release of radon from rocks varies widely. The gas is released from the granites of Cornwall in amounts that can represent a hazard to health if it accumulates in buildings.

The lack of reactivity of the noble gases makes it necessary to examine their physical properties to identify group trends. Table 7.8 gives some essential data for the noble gases.

Table 7.8 *The noble gases.*

Name	Symbol	Electron arrangement	Density (g/dm³)	Boiling point (°C)	Uses
Helium	He	2	180	−269	Balloons, welding
Neon	Ne	2,8	900	−246	Advertising signs
Argon	Ar	2,8,8	1780	−186	Filling light bulbs
Krypton	Kr	2,8,18,8	3744	−153	Specialist lamps
Xenon	Xe	2,8,18,18,8	5890	−108	Specialist lamps
Radon	Rn	2,8,18,32,18,8	9730	−62	Cancer treatment (radioactive)

The transition metals

The transition metals occupy the central part of the Periodic table. As you go across the first transition series from scandium to zinc, one electron is added to the penultimate shell for each element. This is unusual since electrons are usually added to the outermost shell. This means that the transition metals have some distinctive differences from other metals.

The transition metals are:

- very similar in the size of their atoms
- very similar in their properties.

The main differences between them and the other metals of the Periodic table are that they:

- are very dense, with high melting points and boiling points
- can form more than one type of simple ion; for example, iron forms ions with charges +2 and +3
- are often good catalysts; that is, the metals or their compounds can accelerate chemical reactions whilst not forming part of the products
- form brightly coloured compounds
- may be strongly magnetic (iron, cobalt and nickel).

Table 7.9 shows some essential data for some of the transition metals.

Table 7.9 *Data for some of the transition metals.*

Metal	Simple ionic charge	Colour of solution
Chromium(III)	+3	Violet
Manganese(II)	+2	Pale pink
Iron(II)	+2	Green
Iron(III)	+3	Yellow–brown
Copper(II)	+2	Blue or green
Zinc	+2	Colourless (exception)

◆ *Enhancement ideas*

 ◆ Table 7.8 (page 237) is a data table for the noble gases. Use a data book or CD ROM to draw up data tables for the alkali metals (Group 1), the alkaline earth metals (Group 2) and the halogens (Group 7).

 ◆ Use a knowledge of trends in the Periodic table to predict the properties of fluorine, astatine, rubidium and barium. Attempt to make predictions about their physical and chemical properties and check these using data books, CD ROMs or textbooks.

◆ Investigate methods for extracting silver from used photographic fixer by electrolysis or by displacement (using a more reactive metal, such as iron filings).

◆ Place samples of transition metal compounds on a large copy of the Periodic table to show the variety of colours.

◆ Test transition metal samples with a magnet to identify the three metals that are strongly magnetic.

◆ Metals from the first transition series have been responsible for two major periods of technological change in history, the Bronze Age (bronze is an alloy of copper with tin) and the Iron Age. Research the history of the development of bronze and iron.

◆ Pupils can experiment with hydrogen peroxide and a range of metal oxides to find out which metal oxides catalyse its decomposition to give oxygen gas and water. Hydrogen peroxide is corrosive or irritant, depending on its concentration. Some metal oxides are hazardous.

◆ *Other resources*

A range of Periodic tables, Periodic table information and tee-shirts and sweaters featuring the Periodic table are available from Scienceshirts. Contact Gordon Woods, 3 Peterborough Avenue, Oakham, Rutland LE15 6EB.

Data for the elements can be found in *Chemistry Data Book* by Earl and Wilford, 1991, published by Nelson Blackie, ISBN 0 17 438632 X.

Some reactions can be carried out on a very small scale. *Microscale Chemistry* (Royal Society of Chemistry, 1998, ISBN 1 870343 49 2) provides some opportunities. These include:

7 The Periodic table – solubility of sulphates and carbonates of Groups 1 and 2

Class demonstrations of experiments relevant to the Periodic table can be found in *Classic Chemistry Demonstrations* (Royal Society of Chemistry, 1998, ISBN 1 870343 38 7). These are:

ICT materials

The Warwick Spreadsheet System is a range of ICT materials based upon Microsoft Excel. These include the *First Chemistry Pack*. Within this pack there is a spreadsheet investigating trends in properties of the elements within a group and making predictions for the last member. It also produces a variety of plots of densities, melting points and first ionisation energies for the first 56 elements. Other spreadsheets in the pack include ones on acids and alkalis (Chapter 5), alkanes and alcohols (Chapter 3) and rates of reaction (Chapter 8). These are available from Aberdare Publishing Ltd, 6 Nuthurst Grove, Bentley Heath, Solihull B93 8P.

The CD ROM *The Elements* from Yorkshire International Thomson Multimedia Ltd, Television Centre, Kirkstall Road, Leeds LS3 1JS contains a section on the Periodic table. This includes video clips of reactions such as alkali metals with water (including caesium, which could not be carried out in the school laboratory).

Web sites

♦ A short history of the Periodic table and source of references can be found on:
   chemlab.pc.maricopa.edu/periodic/about.html
♦ A Periodic table quiz can be found on:
   www.angelfire.com/hi/BADBOYEE
   This simple quiz, with its errors, may encourage pupils to produce quizzes of their own.
♦ Some interesting alternative periodic tables can be found on:
   chemlab.pc.maricopa.edu/periodic/stowetable.html
   Pupils could be asked to discuss the advantages and disadvantages of each one.

8 | *Rates of reaction*

Philip Hyde

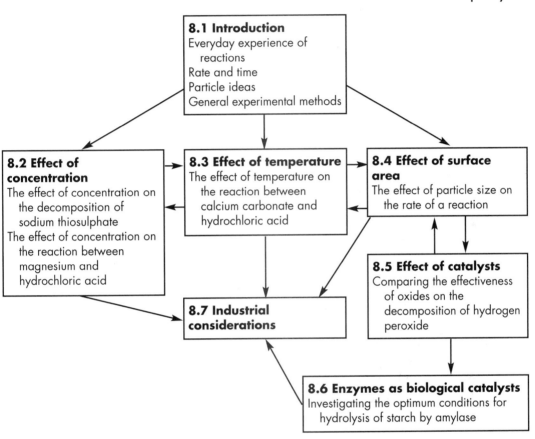

8.1 Introduction
Everyday experience of
    reactions
Rate and time
Particle ideas
General experimental methods

8.2 Effect of concentration
The effect of concentration on
    the decomposition of
    sodium thiosulphate
The effect of concentration on
    the reaction between
    magnesium and
    hydrochloric acid

8.3 Effect of temperature
The effect of temperature on
    the reaction between
    calcium carbonate and
    hydrochloric acid

8.4 Effect of surface area
The effect of particle size on
    the rate of a reaction

8.5 Effect of catalysts
Comparing the effectiveness
    of oxides on the
    decomposition of hydrogen
    peroxide

8.7 Industrial considerations

8.6 Enzymes as biological catalysts
Investigating the optimum conditions for
    hydrolysis of starch by amylase

◆ *Choosing a route*

The study of rates of reaction affords considerable opportunity
for pupils to practise and develop their experimental and
investigative skills. After a suitable introduction, the
experimental work may be attempted in any sequence,
although it would be sensible to leave the enzyme work until
after the basic ideas of catalysis have been met. (Enzymes are
also studied in Biology and it is important to liaise with
Biology teachers to ensure that this topic is taught in a way
which reinforces the concepts without duplication of content.)

The experimental work gives scope for the use of ICT, both
in data-logging and in processing of results, and this can be
used to increase the variety and richness of the experience.

8.1 Introduction

♦ *Previous knowledge and experience*

Pupils may bring some ideas of the particulate nature of matter with them from primary school but mostly these will have been developed in lower secondary school (see Chapter 1).

♦ *A teaching sequence*

Everyday experience of reactions

Pupils will have met a variety of chemical reactions but most will be fairly quick ones in terms of rate. They should be encouraged to think about chemical changes which go on around them and to categorise these into ones which are slow (take days or longer), moderate (take hours or minutes) or fast (take seconds or less). Everyday examples of rates of reaction are given below.

- Slow: rusting and other examples of corrosion, the gradual discoloration of paintwork.
- Moderate: chemical changes involved in cooking.
- Fast: burning reactions (e.g. natural gas, paper and coal), explosives and fireworks, photography.

Rate and time

Having considered that reactions proceed at different speeds, it is then prudent to discuss with pupils the relationship between rate and time. Most should be able to appreciate that a slow reaction takes a long time and that a fast reaction takes a short time, but not all will be able to understand the reciprocal relationship

$$\text{rate} \; \alpha \; \frac{1}{\text{time}}$$

This is needed in the processing of single timing experiments such as those involving sodium thiosulphate (see page 249). Some simple number practice may be useful here to convert times to rates. This is sufficient for most pupils to be able to convert experimental data to workable 'rates', but more advanced pupils may appreciate that the proportionally constant is being ignored to simplify matters.

Particle ideas

For detailed consideration of the necessary particle ideas see Chapter 2. The key concepts which are needed are that:

- reactions take place by chance, when different particles come into contact (collide)
- chemical bonds need to be broken before the particles can rearrange into those of the new substances
- energy is needed to break chemical bonds.

Pupils should appreciate that anything which affects these factors will change the rate of the reaction. It may be appropriate to ask pupils to compile a list of the factors and to predict how the rate will be affected as they are changed.

These ideas should be discussed at this early stage and be reinforced as each experiment is evaluated. Pupils should be encouraged to write in terms of particles, their energies and collisions whenever any discussion of results is attempted.

General experimental methods

Before looking experimentally at the various factors which affect the rate of a chemical reaction, it is useful for the pupils to have a general picture of what has to be done to obtain the necessary numerical data. Guided discussion based on what follows is a quick way to establish this and to enable pupils to contribute to the design of specific experiments, either for normal practical work or for practical assessment purposes.

Simple test-tube demonstrations are useful to provide a focus for the discussion. Suitable examples might be the reactions of:

- marble chips and hydrochloric acid, to illustrate gas evolution
- magnesium and hydrochloric acid, also to illustrate gas evolution
- sodium thiosulphate and hydrochloric acid, to illustrate a reaction involving a precipitate.

To study the rate of a chemical reaction, pupils will need to be able to follow the value of a property of the chemicals which changes during the reaction. The most commonly used reactions for rate studies involve the evolution of a gas and hence can be followed either by collecting the gas and measuring its volume at intervals or by allowing the gas to escape and measuring the loss of mass at intervals. Each of these has a number of variations, the choice being decided by availability of equipment as much as by appropriateness.

Use of an expensive gas syringe (see Figure 8.1a) is normally limited to teacher demonstrations, with a measuring cylinder (see Figure 8.1b) of suitable size being appropriate for use by pupils. The gas syringe can be used to study quite rapid reactions if it is connected to a position sensor as part of a data-logging system (see Figure 8.1c). Remember that a gas syringe has a limited volume (usually 100 cm³), and that quantities need to be determined to suit this.

Figure 8.1
Following the volume of gas evolved during a reaction.
a Using a gas syringe.
b Using a measuring cylinder.
c Using a position sensor and data-logging.

a

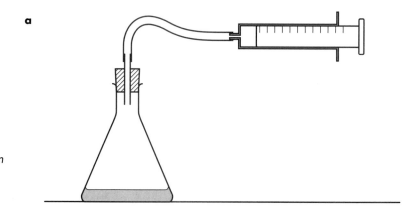

b

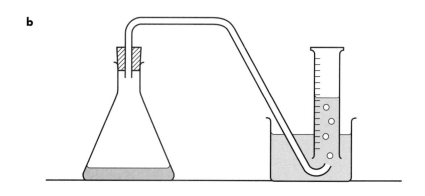

c

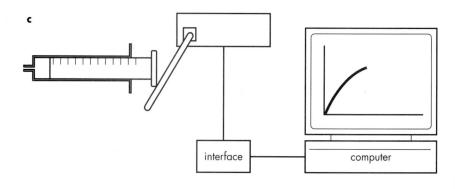

interface | computer

Similarly, schools are unlikely to have sufficient digital balances to allow pupils to use the mass-change apparatus (Figure 8.2a) in small groups, but when it is interfaced to a suitable system with a large-digit display (see Figure 8.2b) pupils can take responsibility for their own mass–time readings and so have greater involvement in the lesson.

Figure 8.2
Following the change in mass during a reaction.
a Using a digital balance.
b Using a sensitive balance interfaced to a large-digit display.

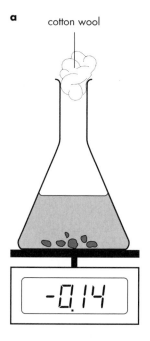

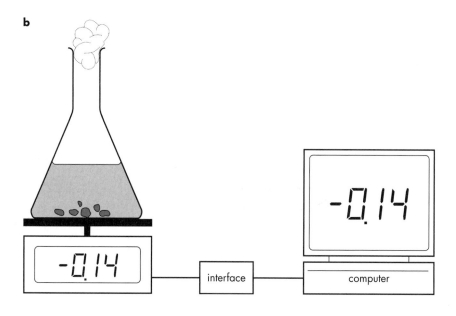

However, it is also possible to study reactions by following the appearance or disappearance of colour using a colorimeter (Figure 8.3a) or by using a light sensor (see Figure 8.3b) as part of a data-logging system. This set-up could also be applicable to reactions which form a precipitate.

Figure 8.3
Following the changes in colour or turbidity during a reaction.
a Using a colorimeter.
b Using a light sensor and data-logging.

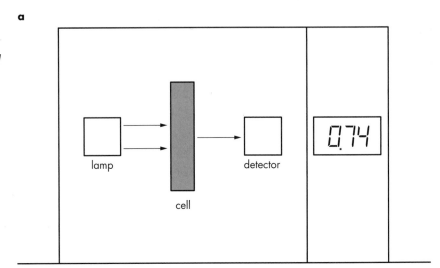

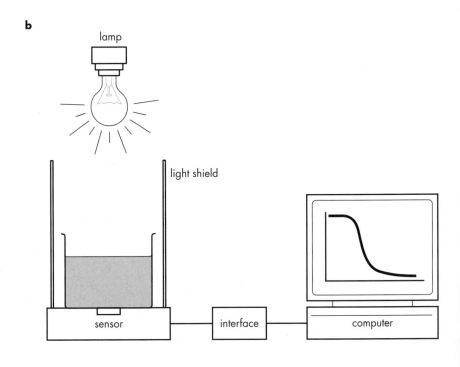

Suggestions are given in the appropriate experimental sections that follow, but you should be willing to try out other variations as you gain in confidence.

Pupils also need to be given some guidance regarding the interpretation of reaction curves, which are typically of the form seen in Figure 8.4.

Figure 8.4
A typical reaction curve.

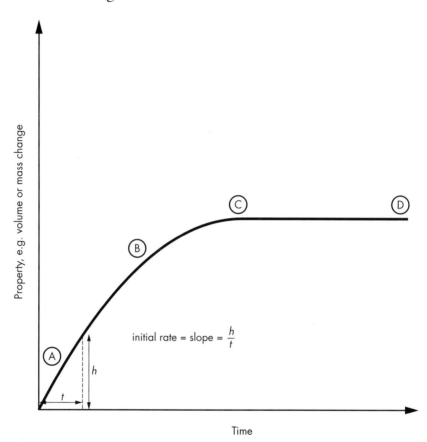

They need to be aware that any comparison between the rates of different experiments should be done at the early, linear section of the graph (A), when reagent concentrations have not fallen significantly. Qualitative comparison can be done simply by comparing the steepness of these sections, but quantitative work needs the initial rate to be calculated from the slope as shown in the diagram. This is essential if the work is to be extended to investigate the exact relationship between a factor and its effect on the rate. It is worth noting that it is not necessary to follow a reaction to completion if the aim is to compare rates: it is sufficient to take enough readings to establish the straight-line section. This saves time and enables more experiments to be done in any one practical session. Pupils should also be aware that the curved section (B) is due to depletion of reagents, and that the reaction has finished at the *first* point where the curve levels off (C) and not where the graph finishes (D).

To show the relationship between the rate and a factor which affects it, e.g. the concentration, several experiments need to be carried out at different concentrations and the rates calculated for each concentration (see Figure 8.5a). The values of rates and concentrations are plotted on a separate graph (Figure 8.5b) to show the relationship (in this case, proportional).

Figure 8.5
Investigating the relationship between concentration and rate.
a Using reaction curves to determine the initial rate at five different concentrations.
b Plotting the initial rates against concentration shows that rate is proportional to concentration in this case.

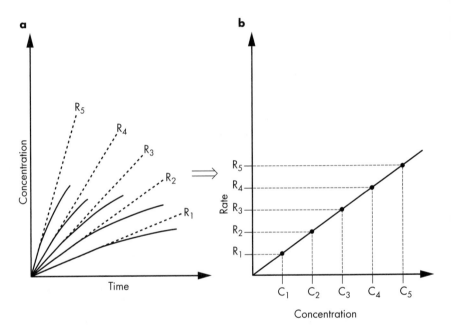

◆ *Enhancement ideas*

◆ Pupils could compile a scrapbook of examples of rates of reaction. For example explosions in flour mills or mines are very rapid reactions, erosion of rocks and buildings are very slow reactions, etc.

8.2 Effect of concentration

◆ *Previous knowledge and experience*

From lower secondary school, pupils may have the idea that reactions are faster when the reacting solutions are more concentrated.

◆ *A teaching sequence*

The idea of reactions being faster when reacting solutions are more concentrated can be refreshed by getting pupils to start with 2 mol/dm^3 hydrochloric acid and dilute it with equal volumes of water to get 1 mol/dm^3, 0.5 mol/dm^3 and 0.25 mol/dm^3 acid. Then add a piece of magnesium to each solution. (Note that magnesium is highly flammable.) The time taken for each piece to react should be recorded. The time gets longer as the solutions become more dilute.

The effect of concentration on the decomposition of sodium thiosulphate

Pupils need to be aware that as the concentration increases the number of reacting particles in a given volume increases, resulting in a greater rate of collision and hence a greater chance of reaction (see Figure 8.6). It is not too difficult to predict that the rate of reaction should increase as concentration increases, so the purpose of the experimental work should be to establish the *relationship* between the two.

Figure 8.6
The effect of concentration on the probability of a collision between particles of A and X.
a *At low concentrations of A and X there is little chance of a collision.*
b *At higher concentrations, a collision is more likely.*

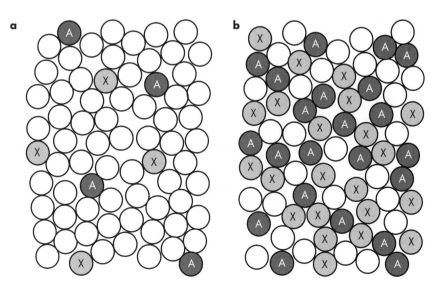

Sodium thiosulphate is stable in aqueous solution but decomposes rapidly in the presence of acid:

$$Na_2S_2O_3 \text{ (aq)} + 2HCl \text{ (aq)} \rightarrow 2NaCl\text{(aq)} + H_2O\text{(l)} + SO_2\text{(g)} + S\text{(s)}$$

A simple way for pupils to investigate the reaction is to follow the appearance of the sulphur precipitate.

Materials
- eye protection
- sodium thiosulphate solution (0.2 mol/dm³)
- hydrochloric acid solution (0.2 mol/dm³)
- distilled water
- 3 small measuring cylinders (e.g. 25 cm³)
- stopwatch
- white paper marked with bold black cross
- 50 cm³ tall beaker

Safety
- *Sulphur dioxide is toxic and corrosive; it may also trigger an asthma attack in a sensitive individual. Pupils with asthma should be warned to stay away from the solution.*
- *Sodium thiosulphate is not harmful unless ingested in large amounts.*
- *Pour the experimental solution away as soon as it has been used. Preferably use a sink in a fume cupboard. If a sink in the open lab is used, then flush with plenty of cold water.*
- *Wear eye protection when using hydrochloric acid.*

Procedure
1. Measure 10 cm³ of the thiosulphate solution into the beaker, using a measuring cylinder.
2. Place the beaker on to the paper marked with the cross.
3. Measure 10 cm³ of the hydrochloric acid into a measuring cylinder.
4. Add the acid to the thiosulphate, start the stopwatch and mix the solutions well.
5. Note the time taken for enough sulphur to be precipitated to mask the cross.
6. Repeat the experiment several times:
   a) gradually diluting the acid with water, e.g. 9 cm³ acid and 1 cm³ water, 8 cm³ acid and 2 cm³ water, etc.
   b) gradually diluting the sodium thiosulphate with water in the same way.

What you might expect

Table 8.1 *The effect of acid concentration on the decomposition of sodium thiosulphate.*

Volume of sodium thiosulphate (cm³)	Volume of hydrochloric acid (cm³)	Volume of water (cm³)	Time for cross to disappear (s)	Rate = 1/time (s⁻¹)
10	1	9	588	0.0017
10	2	8	370	0.0027
10	3	7	212	0.0047
10	4	6	178	0.0056
10	5	5	135	0.0074
10	6	4	112	0.0089
10	7	3	99	0.0101
10	8	2	85	0.0118
10	9	1	77	0.0130
10	10	0	70	0.0143

Table 8.1 shows a set of sample results. As the concentration of acid or thiosulphate is reduced, the reaction time becomes longer and can be plotted as a simple line graph, with the shape shown in Figure 8.7a, overleaf. It is desirable that more advanced pupils attempt to use the reciprocal relationship between time and rate as well as working out the actual concentrations of reagent in each solution to produce a graph such as in Figure 8.7b, which shows the expected proportional relationship.

In terms of class management, pupils can be divided into groups looking at different reagents, or even individual mixtures, in order to gather and discuss results. Best results are obtained when the controlled reagent is kept well in excess. Some practical difficulty might be found in the exact determination of the time when the cross is masked, especially for the slower reactions. This is a good point to use as a basis for discussing experimental errors with suitable pupils.

RATES OF REACTION

Figure 8.7
Determining a rate relationship from timing experiments.
a *A simple plot of the timing results.*
b *An inverse plot shows that the rate is proportional to the concentration of acid.*

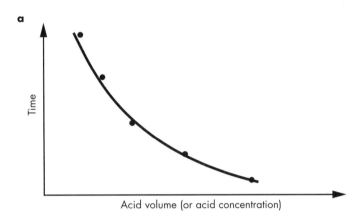

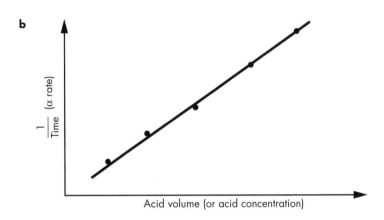

One important aspect is the 'single timing' nature of the experiment. Where is the timing point on the reaction curve? If the experiment is done using a computer to follow the change in light level as sulphur forms (Figure 8.3b, page 246), then a curve such as the one shown in Figure 8.8 is produced. The use of the computer can be demonstrated and pupils can analyse sets of graphs produced previously in order to select a good timing point. Reference to the previous experiment may be a useful starting point for ideas. Some examples of points to choose might be:

- the time up to the first noticeable loss in light (A)
- the time to reach a certain light level, e.g. 90% or 80%, which might correspond to the covering of the cross (B)
- the time taken to reach zero (or the limit of) light transmission (C).

Provided there is little delay between mixing and the start of recording, the use of early points tends to give better correlation.

Figure 8.8
Changes in light transmission as sulphur is precipitated.

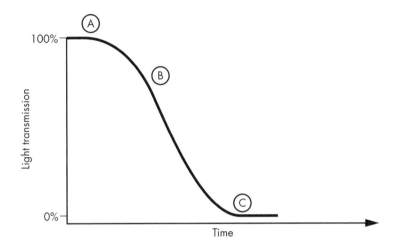

When using the computer as a data-logger, first calibrate the system. This is done following the procedure appropriate to your equipment, setting a value of zero with the light off and a value of 100% transmission with the light on and distilled water in the apparatus. Once this has been done, follow the method above but start recording as the solutions are mixed.

The effect of concentration on the reaction between magnesium and hydrochloric acid

$$Mg(s) + 2HCl(aq) \rightarrow MgCl_2(aq) + H_2(g)$$

The reaction is quite rapid and so is suitable to use at low acid concentrations.

Materials
- eye protection
- hydrochloric acid (1 mol/dm^3)
- magnesium ribbon
- distilled water
- conical flask
- delivery tube
- water trough
- measuring cylinders to collect gas, 100 or 250 cm^3
- stopwatch

Safety
- *Magnesium is highly flammable.*
- *1 mol/dm^3 hydrochloric acid is low hazard.*
- *Limit pupils to 0.1 g of magnesium, which will give 100 cm^3 of gas.*
- *Wear eye protection when using hydrochloric acid because of acid spray when bubbles burst.*

RATES OF REACTION

Procedure

1. Measure out 50 cm^3 of 1 mol/dm^3 hydrochloric acid and place it into a conical flask.
2. Measure 0.1 g of magnesium ribbon.
3. Set up the apparatus as in Figure 8.1b (page 244).
4. Add the magnesium to the acid, stopper quickly and begin timing.
5. Take volume readings at convenient intervals (10–30 s).
6. Repeat several times, each time replacing some acid by water, e.g. 45 cm^3 acid and 5 cm^3 water, then 40 cm^3 acid and 10 cm^3 water, etc. (More advanced pupils should calculate the actual concentrations.)
7. Plot volume–time graphs to establish the rate for each concentration.
8. Plot a rate–concentration graph to establish the (proportional) relationship.

What you might expect

Table 8.2 shows sets of volume–time results for a range of acid concentrations.

Table 8.2 *The effect of acid concentration in the magnesium/hydrochloric acid reaction.*

Concentration of acid (mol/dm³)	Volume of gas (cm³) at the time shown (s)									
	0	20	40	60	80	100	120	140	160	180
1.0	0	6	14	20	28	33	40	46	52	58
0.9	0	5	11	18	22	28	35	40	44	50
0.8	0	5	10	14	21	27	32	36	43	48
0.7	0	4	8	15	19	23	27	33	38	43
0.6	0	4	7	13	16	19	24	29	31	35
0.5	0	2	6	10	13	17	19	23	26	30

Plotting these produces the initial rates of reaction listed in Table 8.3.

Table 8.3 *Quantitative effect of acid concentration on rate of the magnesium/hydrochloric acid reaction.*

Concentration of acid (mol/dm³)	1.0	0.9	0.8	0.7	0.6	0.5
Rate (cm³/s)	0.335	0.285	0.265	0.230	0.195	0.165

The speed of the reaction will lessen as more water is added but, if the reaction is taken far enough, the total amount of gas should not change, as the acid remains in excess.

The rate–concentration graph should be a good straight line, passing through the origin, confirming the proportionality between the two.

At the conclusion of the experiment, pupils can be asked to evaluate the process. The following points could be considered.

- How many concentrations would be suitable to establish a reliable relationship?
- How accurate are the dilutions?
- How reproducible are the rate values for a given concentration?
- How important is the delay between adding the magnesium, sealing the flask and starting the timing?
- Is the lack of mixing important?

An equivalent experiment can be carried out using marble chips and hydrochloric acid. The reaction is slower and 1–2 g of fine marble chips can replace the magnesium, but the process remains the same. (Other practical detail follows on page 261.)

♦ *Enhancement ideas*

- ♦ The use of a rate–concentration graph as a calibration curve to establish an unknown concentration by measuring a rate. This could be applied quite easily to reactions involving acids but might also be applied in a rate context by studying old samples of hydrogen peroxide to see how much their concentrations have changed and hence estimate the speed of its thermal (as opposed to catalytic) decomposition.

- ♦ Pupils could look at the 'iodine clock' reaction (that is the basis of the 'water to ink' magic trick involving the mixture of potassium iodide, sodium thiosulphate and starch solution, added to sulphuric acid and hydrogen peroxide). Pupils could be asked to work out a reliable mixture of the various chemicals to give a precise time delay so that a magician could perfect a routine. Please note that at the concentrations required for this reaction, only hydrogen peroxide and sulphuric acid present a low hazard and should be treated with care. It is important to rinse away all solutions with plenty of water at the end of the experiment.

8.3 Effect of temperature

◆ *Previous knowledge and experience*

Again pupils may expect reactions to go faster at higher temperatures. Iron hardly reacts with dilute hydrochloric acid at room temperature but reacts faster when the acid is heated.

◆ *A teaching sequence*

Pupils need to be aware that at higher temperatures:

- the energy of the particles is increased
- collisions occur with greater energy
- bond breaking is therefore more likely in any given collision
- the chance of reaction is increased.

The concept of 'activation energy' is now confined largely to post-16 syllabuses and need not be discussed, except with suitably advanced pupils. However, there is a common misconception that the increase of rate with temperature is mainly due to the increased frequency of collisions. This is not the case. A 10 °C rise in temperature does produce a few per cent increase in the collision frequency but normally results in about a doubling of the reaction rate. The discrepancy between the two can only be accounted for by the increase in the percentage of collisions that lead to a reaction.

Pupils need to appreciate that this doubling goes on for *each* 10 °C rise; a change from 20 °C to 100 °C is 8×10 °C and will produce an approximate rate increase of $2^8 = 256$ times. This is of great practical importance. For example, if a pressure cooker can operate at 110 °C rather than an open pan boiling at 100 °C, the cooking speed is doubled and the cooking time is halved. The high temperatures which are achievable in industry may result in reaction rates which are many thousands, hundreds of thousands or even millions of times greater than those which are possible at room temperature and thus, despite the extra expenditure of energy, become economic. Pupils could work out for themselves the rate increase between room temperature and the operating temperature of an important industrial process (see Chapter 9).

The effect of temperature on the reaction between calcium carbonate and hydrochloric acid

<u>Materials</u>
- eye protection
- hydrochloric acid (1 mol/dm^3)
- marble pieces, carefully graded to an even size and acid-washed to remove surface powder
- conical flask
- delivery tube
- water trough
- 250 cm^3 measuring cylinder
- stopwatch
- ice
- Bunsen burner, tripod, gauze and heatproof mat
- thermometer, (-10 to $110\,°C$)

<u>Safety</u>
- *1 mol/dm^3 hydrochloric acid is low hazard, but should be treated with care at elevated temperatures.*
- *An upper temperature limit of 50 °C is advisable to minimise the release of acidic fumes.*
- *Wear eye protection when using hydrochloric acid because of acid spray when bubbles burst.*

<u>Procedure</u>
1. Measure out 2 g of the marble pieces.
2. Measure 50 cm^3 of the hydrochloric acid into the flask and note its temperature.
3. Assemble the apparatus as in Figure 8.1b (page 244).
4. Add the marble pieces to the acid, seal the apparatus and begin taking volume–time readings at convenient intervals (30–60 s).
5. Repeat several times, changing the temperature by cooling the acid in an ice–water mixture or by heating it.
6. Plot volume–time graphs to find the rate at each temperature.
7. Plot a rate–temperature graph.

<u>What you might expect</u>
Table 8.4 (overleaf) shows sets of volume–time results at different temperatures. Plotting these produces the initial rates listed in Table 8.5.

Table 8.4 *Effect of temperature on the calcium carbonate/hydrochloric acid reaction.*

1 °C	Time (s)	0	30	60	90	120	150	180
	Volume (cm³)	0	16	30	47	63	77	96
10 °C	Time (s)	0	30	60	90	120	150	180
	Volume (cm³)	0	22	48	74	95	125	152
19 °C	Time (s)	0	30	60	90	120	150	180
	Volume (cm³)	0	34	70	110	140	171	196
30 °C	Time (s)	0	20	40	60	80	100	120
	Volume (cm³)	0	46	81	124	151	179	205
40 °C	Time (s)	0	10	20	30	40	50	60
	Volume (cm³)	0	32	75	109	135	162	190
50 °C	Time (s)	0	10	20	30	40	50	60
	Volume (cm³)	0	50	98	148	180	214	236

Table 8.5 *Temperature/rate data for the calcium carbonate/hydrochloric acid reaction.*

Temperature (°C)	1	10	19	30	40	50
Rate (cm³)	0.50	0.80	1.17	2.11	3.38	4.93

Pupils will find that the rate increases with temperature but that the relationship is not proportional. In fact, the rate increases exponentially with temperature (see Figure 8.9). This should add weight to the idea that it is the *energy* of the collision which is important.

Figure 8.9
The rate of reaction increases exponentially with temperature.

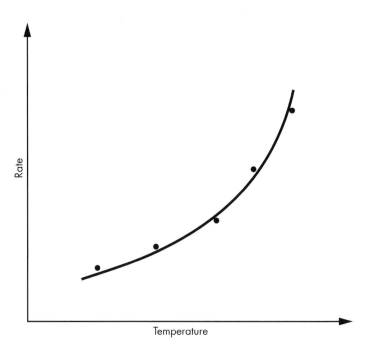

Some discussion points which might be relevant to the evaluation are the following.

- The temperature will not have been constant during the whole of the experiment. How could this be improved? (For example, by using thermostatic baths.)
- The reaction is exothermic. Would this add to the problem of temperature control?
- How many different temperatures would be suitable to establish a reliable relationship?
- Is the Celsius scale the most appropriate? (The Kelvin scale should really be used, especially if more advanced pupils wish to look more scientifically into the relationship between rate and temperature by using logarithmic graphs. The actual relationship is $\log(\text{rate}) \propto 1/T$, and the activation energy is obtained from the slope of the graph – refer to advanced textbooks.)

◆ *Further activities*

- ◆ Substitute magnesium for the marble chips, as in the concentration experiment (page 253).

- ◆ Repeat the thiosulphate experiment over a range of temperatures from 1 °C to 50 °C (page 250). Great care will be needed as sulphur dioxide will be less soluble at higher temperatures. Cool solutions in a fridge so you can work below room temperature. This should ensure you can get five readings without going to too high a temperature.

- ◆ Use the mass method (Figure 8.2, page 245) instead of the volume method. For details of this, see page 261.

8.4 Effect of surface area

◆ *Previous knowledge and experience*

Pupils may be aware that substances react faster when they are powdered. They may have seen, for example, the burning of iron filings and a nail when heated in air.

◆ *A teaching sequence*

Pupils need to be aware that when one of the reagents is a solid and the others are in solution, reaction can only take place on the surface of the solid. The reaction rate should therefore increase in proportion to the available surface. This can be reinforced by diagram or calculation. Small wooden cubes (Tillich blocks) can be a useful aid here (see Figure 8.10). The difficulty is in actually quantifying the change in surface area, but it is easy to show the qualitative effect.

Figure 8.10
Using wooden blocks to show how particle size affects the surface area.
a A single cube of side 2 cm has a total surface area of
$6 \times 2 \times 2 = 24$ *cm^2.*
b If the cube is divided into eight 1 cm cubes, the surface area is now
$8 \times (6 \times 1 \times 1) = 48$ *cm^2.*

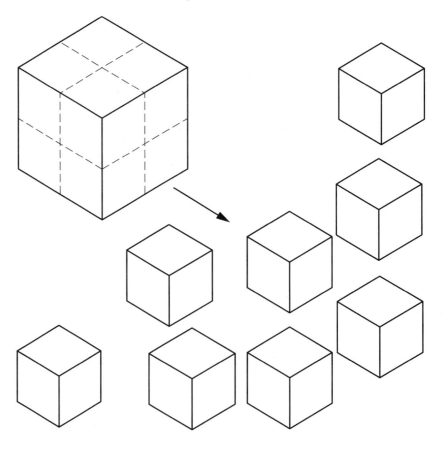

The effect of particle size on the rate of a reaction

Materials
- eye protection
- hydrochloric acid (1 mol/dm³)
- powdered calcium carbonate
- large, medium and small pieces of marble
- conical flask
- access to balance
- boiling tubes

Safety
- *1 mol/dm³ hydrochloric acid is low hazard.*
- *The reaction involving powdered calcium carbonate should be carried out over a sink because of the spray.*
- *Wear eye protection when using hydrochloric acid because of acid spray when bubbles burst.*

Qualitative procedure 1
1. Add a 1 g piece of marble to 10 cm³ of hydrochloric acid.
2. Repeat with 1 g of powder and fresh acid. Note that the powder produces considerable effervescence, indicating a much faster reaction.

Qualitative procedure 2
1. Sort the marble pieces into sizes as follows:
large: approximately 0.5 cm
medium: approximately 0.25 cm
small: approximately 0.1 cm
2. Measure 50 cm³ of hydrochloric acid into the flask. Add 2 g sample of large marble pieces.
3. Quickly assemble the apparatus as in Figure 8.2a (page 245).
4. Take mass readings at regular intervals.
5. Repeat with fresh samples of acid and each of the other samples of marble. Carry out separately at room temperature for 2 g of each of the sizes of marble pieces.

What you might expect
Table 8.6 shows a set of sample results, showing the mass loss (in grams) with time for different sized pieces.

Table 8.6 *Mass loss from different size marble pieces in reaction with hydrochloric acid.*

Time (min)	0	2	4	8	12	16	20
Small	0.00	0.13	0.25	0.52	0.78	0.97	1.15
Medium	0.00	0.08	0.15	0.29	0.45	0.61	0.75
Large	0.00	0.06	0.12	0.25	0.41	0.53	0.68

As the pieces get smaller, the reaction rate increases significantly. Depending on the actual sizes used, the small pieces should react several times faster than the large ones. For the results in Table 8.6, the rates are:

small	0.065 g/min
medium	0.038 g/min
large	0.031 g/min

This approach should be sufficient for the vast majority of pupils, but those with good maths skills may benefit from extending the investigation to try to estimate the surface area and relate it to the increase in rate.

<u>Semi-quantitative procedure</u>

1. Sort out the marble pieces into several 2 g amounts, consisting of different numbers of pieces as shown, with the pieces in each group being approximately the same size, so as to make up the total mass fairly accurately.

1 piece
2 pieces
4 pieces
8 pieces
16 pieces
32 pieces

2. Carry out the same experiment as described above for each piece size and find the rate from each graph.

3. Estimate the surface area of marble in the different samples as follows. Given that the density of marble is 2.7 g/cm³, it is possible to calculate that 2 g should have a volume of 0.741 cm³. This can then be approximated as either a cube of side $l = 0.905$ cm (volume $= l^3$) or a sphere of radius $r = 0.561$ cm (volume $= \frac{4}{3}\pi r^3$). Pupils can then use standard formulae to calculate the relevant surface area (area of a cube $= 6 \times l^2$; area of a sphere $= 4\pi r^2$). The calculation then needs to be extended for the relevant number of smaller cubes or spheres of the same total volume. This is an ideal opportunity to use a spreadsheet and explore the effects of different approximations. Different groups of pupils could look at different approximations and report back to the class.

4. Each method of estimating the surface area should be correlated with the rate by plotting a graph of rate against estimated surface area.

<u>What you might expect</u>
Table 8.7 shows a set of sample results, with the areas estimated using the cube approximation.

Table 8.7 *Correlation of rates with approximated surface area.*

Number of pieces	Estimated surface area (cm²)	Rate (g/min)
1	4.91	0.0416
2	6.19	0.0516
4	7.80	0.0760
8	9.82	0.0886
16	12.38	0.1120
32	15.60	0.1400

There is a good chance of finding a proportional relationship between surface area and rate but the correlation might not be as convincing as that usually found between concentration and rate. The marble pieces are certainly not cubes or spheres, and in any given experiment they are not all the same size.

One variation might be to carry out the investigation using magnesium instead of marble. If a magnesium strip is cut into successively smaller pieces for each experiment, the small increase in surface area can be calculated quite accurately after initial measurement of the dimensions of the strip. Care needs to be taken to get the best accuracy from the rate measurements for a good correlation. Magnesium is highly flammable. Wear eye protection when using hydrochloric acid because of acid spray when bubbles burst.

Please note that to detect a loss of mass from magnesium and hydrochloric acid may be problematic for pupils. 100 cm³ of hydrogen at room temperature has a mass of approximately 0.009 g – too small to be significant on most balances. However, this experiment should be seen as an opportunity for pupils to recognise that despite careful planning and measurement not all scientific investigations provide the expected results, and even in these instances conclusions can still be drawn.

8.5 Effect of catalysts

♦ *Previous knowledge and experience*

Pupils are likely to have heard of catalytic converters, though they are unlikely to know any details of their function. In Biology lessons, they may have come across enzymes as examples of catalysts.

♦ *A teaching sequence*

Catalysts are substances which increase the rate of a chemical reaction without being chemically changed themselves. Catalysts actually provide an alternative pathway for the reaction with a lower activation energy, which results in a greater fraction of collisions having the necessary energy for reaction at a given temperature. Less advanced pupils may be satisfied with an explanation which has the catalyst 'making it easier to break bonds'.

Some common misconceptions are that

- the catalyst does not get involved: it must have *some* involvement in order to affect the reaction!
- there is a bottle marked 'catalyst' containing a substance which can be brought out to be added to any reaction which needs speeding up: pupils need to understand that catalysts have to be found by experiment for each reaction and that a substance which is very good for one reaction may be totally ineffective for another
- no more is produced using a catalyst: the same amount is produced, but more quickly.

Pupils should be made aware that

- many catalysts are transition metals or their compounds (see *Industrial considerations*, page 271)
- catalysts are usually needed in small amounts compared to the amounts of the reactants
- the catalyst should be recoverable, with its mass unchanged
- catalysts do not operate indefinitely: in practice, they may be rendered inactive (poisoned) by impurities or be broken down mechanically.

Comparing the effectiveness of oxides on the decomposition of hydrogen peroxide

Hydrogen peroxide decomposes according to the equation

$$2H_2O_2(aq) \rightarrow 2H_2O(l) + O_2(g)$$

The decomposition is quite slow at room temperature and can be regarded as nil over the course of a normal lesson session.

Materials
- 10 volume hydrogen peroxide
- manganese(IV) oxide
- magnesium oxide
- lead(II) oxide
- lead(IV) oxide
- conical flask
- delivery tube
- gas syringe
- stopwatch
- measuring cylinders

10 volume hydrogen peroxide decomposes to produce $10 \ cm^3$ of oxygen from each $1 \ cm^3$ of hydrogen peroxide. It is more economical to buy hydrogen peroxide in higher concentrations. This should then be diluted and stored in the diluted form as it decomposes more slowly.

Safety
- *10 volume hydrogen peroxide is of low hazard; concentrated hydrogen peroxide is corrosive and oxidising.*
- *Manganese(IV) oxide is harmful by inhalation and by swallowing.*
- *Lead(II) oxide and lead(IV) oxide are toxic by inhalation and by swallowing.*
- *Lead compounds have a danger of cumulative effects and may harm an unborn child.*

Procedure
1. Measure out $50 \ cm^3$ of water into the conical flask.
2. Disperse 0.1 g of the manganese(IV) oxide into the water.
3. Assemble the delivery tube and gas syringe as in Figure 8.1a (page 244).
4. Add $10 \ cm^3$ of the hydrogen peroxide and seal the flask quickly.
5. Take volume–time readings at convenient intervals, gently agitating the reaction flask.
6. Repeat for each oxide.
7. Plot volume–time curves and determine the rates of reaction.

<u>What you might expect</u>

Table 8.8 shows a set of sample results for MnO_2 and PbO_2, giving the volumes of gas produced with each catalyst, in cm^3.

Table 8.8 *Effect of different metal oxides on decomposition of hydrogen peroxide.*

Time (s)	0	15	30	45	60	75	90	105	120	135	150
MnO_2 catalyst	0	2	6	8	10	13	16	18	20	24	26
PbO_2 catalyst	0	11	24	35	45	57	64	71	76	80	85

Magnesium oxide and lead(II) oxide have little or no effect on the reaction, manganese(IV) oxide increases the rate significantly (to $0.172 \ cm^3/s$ here), but lead(IV) oxide increases it still further (to $0.77 \ cm^3/s$).

Evaluation of the experiment could consider the following points.

• The masses of oxides are the same, but are the powders the same 'fineness'?
• Are the oxides which increase the rate actually catalysts? (This could be checked by recovering them on a filter paper, drying and re-weighing.)
• Is the mixing important? (Use of a magnetic stirrer may aid reproducibility of results.)

♦ *Further activities*

 ♦ This experiment is ideal for data-logging using apparatus such as that in Figure 8.1c (page 244). The position sensor needs to be calibrated so that one extreme reads zero volume and the other reads the actual maximum volume which can be recorded at the end of its travel (usually about $80 \ cm^3$). A magnetic stirrer is desirable for comparability of mixing, and it is useful if the software can display several runs graphically for immediate comparison. Ensure that the gas syringe is perfectly clean so that it runs freely, otherwise it will tend to stick at slow rates of gas evolution.

8.6 Enzymes as biological catalysts

◆ *Previous knowledge and experience*

Pupils may have come across enzymes in Biology or Home Economics lessons. They may be aware that 'biological' soap powders contain enzymes.

◆ *A teaching sequence*

Enzymes are large protein molecules which act as catalysts in biological systems. They are very specific in their action, usually acting only on one type of molecule and carrying out one type of reaction.

Enzymes are different from other catalysts in that their reactions show optimum rates as temperature and pH are raised. Their specificity is usually interpreted in terms of a 'lock and key' model (see Figure 8.11), in which the reactant molecule fits into a site in the complex protein structure, weakening bonds and allowing the reaction to take place more easily. The protein structure and the shape of the active site are affected by changes in pH (usually reversibly) and by high temperatures, which permanently denature the protein (as in the conversion of raw egg white into the cooked form). In their writing, lower ability pupils often refer to enzymes being 'killed', which shows a misunderstanding of their nature that should be corrected.

Figure 8.11
The 'lock and key' model of enzyme reaction.

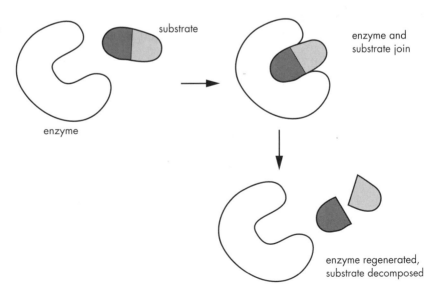

substrate

enzyme and substrate join

enzyme

enzyme regenerated, substrate decomposed

One easily studied reaction is the hydrolysis of starch to glucose by the action of amylase:

$$\text{starch} \xrightarrow{\text{amylase}} \text{glucose}$$

The conversion of the starch can be followed by extracting samples periodically from the mixture and testing them with iodine solution; this gives a blue-black colour with any starch present.

Investigating the optimum conditions for hydrolysis of starch by amylase

Materials
- eye protection
- soluble starch solution (1%)
- amylase solution (1%)
- ethanoic acid (0.1 mol/dm^3)
- sodium carbonate solution (0.05 mol/dm^3)
- dropping pipettes
- test tiles
- iodine solution
- water baths at different temperatures
- ice
- thermometers
- boiling tubes

Safety
- *Amylase powder is an irritant and potential allergen. Use of disposable gloves is a sensible precaution for the technician preparing the solution.*
- *0.1 mol/dm^3 ethanoic acid is irritant.*
- *Iodine solution is harmful above a concentration of 1 mol/dm^3.*
- *Iodine is harmful by inhalation and skin contact and causes burns over time.*
- *Iodine vapour crystallises very painfully on eyeballs.*
- *Wear eye protection when dispensing solutions.*

Procedure
1. Measure out 20 cm³ of starch solution at room temperature into a boiling tube. Add 3 cm³ of water.
2. Add 2 cm² of the amylase, start timing and mix well.
3. Remove 0.5 cm³ at convenient intervals on to a test tile and add one or two drops of iodine.
4. Continue this until no blue–black colour is seen on adding the iodine. Note this time.
5. Repeat at different temperatures between 0 °C and 60 °C. (The reaction may also be attempted in boiling water.)
6. Repeat, replacing small amounts of the water by acid or alkali, e.g. 2.5 cm³ water and 0.5 cm³ acid, 2 cm³ water and 1 cm³ acid, etc.

What you might expect
Table 8.9 and Table 8.10 show sample results to illustrate the effects of temperature and pH, respectively.

Table 8.9   *The effect of temperature on the enzyme-catalysed hydrolysis of starch.*

Temperature (°C)	1	10	19	31	38	50	60	100
Reaction time (min)	>60	13	4.5	3	2	15	>60	∞

Table 8.10   *The effect of pH on the enzyme-catalysed hydrolysis of starch.*

Added substance	Reaction time (min)
1.0 cm³ ethanoic acid	>15
0.5 cm³ ethanoic acid	8.5
Nil	3
0.5 cm³ sodium carbonate	2.25
1.0 cm³ sodium carbonate	>15

As the temperature rises from 0 °C, the time taken to convert the starch decreases as the rate increases, in agreement with the previous study of the effect of temperature. Above 40 °C, however, the time begins to lengthen once more as the protein structure becomes more and more denatured and the rate of reaction lessens. Optimum temperatures for digestive enzymes are around normal body temperature (37 °C).

Amylase has an optimum pH of around 8. A small amount of alkali may be seen to increase the rate a little, but acid and larger amounts of alkali will decrease it, giving longer reaction times. More advanced pupils could determine the exact pH using a pH meter and/or carry out the reaction in buffer solutions. (Most laboratory handbooks give recipes for buffers in the range from pH 5 to pH 8; these are based on sodium and potassium hydrogenphosphates and are quick and easy to make up.)

The time for the starch to be removed can be difficult to decide. The blue–black colour is replaced by a dark red–brown colour due to a dextrin–iodine complex which persists much longer than the starch. If pupils wait until *no* coloration is seen, reaction times will be much longer than indicated. If tests are done in sequence on the test tile, visual inspection can be done at leisure to decide when the absence of the blue–black complex first occurred. Reagents must be freshly made for best results, but the starch and amylase solutions can be stored in a refrigerator for a few days if necessary.

The detergent industry has developed enzymes that have maximum activities well outside the normal ranges of temperatures and pH, e.g. at 80 °C and pH 9.

◆ *Enhancement ideas*

◆ Pupils could look at the effect of conditions on the time needed to make yoghurt. Fine control over temperature and pH would be needed as well as some appreciation of the food safety aspects of any additives.

◆ The use of enzymes in washing powders could be investigated in various ways, for example by using a standard soiled cloth each time and looking at the separate effects of immersion time and temperature on the cleaning ability of the powder. An alternative to the soiled sample would be to use standard solutions of albumin and starch to investigate how well specific food groups are digested in the cleaning process. A range of different powders could be compared under the same conditions. If available, old samples could be compared with new samples to see how quickly the enzymes lose their activity.

8.7 Industrial considerations

◆ *Previous knowledge and experience*

Pupils are unlikely to have come across the use of catalysts in industrial processes.

◆ *A teaching sequence*

Industry requires the largest amount of product, in the shortest time and at the lowest cost. Catalysts are often the key factor in making a reaction work and in making it perform economically. Many of these reactions are reversible and the high temperatures used actually reduce the yield of product, but a compromise temperature is chosen which gives sufficient yield in a proportionately shorter time (see Chapter 9).

Some examples which could be researched by pupils include:

- the production of ammonia in the Haber process (iron catalyst)
- the oxidation of ammonia over a platinum catalyst in the production of nitric acid
- the key reaction in the Contact process, i.e. the conversion of sulphur dioxide into sulphur trioxide using vanadium(V) oxide
- cracking of the naphtha fraction from petroleum distillation using an aluminium oxide catalyst
- the production of margarines by hydrogenation of vegetable oil using a nickel catalyst.

Similarly, pupils could examine the use of enzymes in:

- the brewing industry
- cheese manufacture
- biological washing powders
- the body, for building molecules as well as breaking them down.

More detailed information can be found in the relevant section of the companion Biology book. Topical information could be found using an internet search.

♦ *Other resources*

Classic Chemistry Demonstrations (Royal Society of Chemistry, 1998, ISBN 1 870343 38 7) provides a range of experiments or demonstrations relevant to this section, as detailed below.

3   A visible activated complex

4   An oscillating reaction

12   Catalysis of the reaction between sodium thiosulphate and hydrogen peroxide

23   The 'Old Nassau' clock reaction

45   A solid–solid reaction

58   Catalysts for the decomposition of hydrogen peroxide

62   The spontaneous combustion of iron

78   Following the reaction of sodium thiosulphate and acid using a colorimeter

88   Catalysts for the thermal decomposition of potassium chlorate

99   The cornflour bomb

100   The oxidation of ammonia

The effect of concentration on the rate of a reaction can be shown by pupils on a very small scale using experiment 23 in *Microscale Chemistry* (RSC, 1998, ISBN 1 870343 49 2).

More information on the use of enzymes in the dairy industry can be found in *Applied Microbiology in the Dairy Industry* in Hobson's Science Support series (1989), ISBN 1 85324 369 8.

ICT materials

Experiments on rates of reaction provide good opportunities for investigations and for pupils to demonstrate their use of ICT. Examination boards will accept investigations that are word processed and the use of graph packages. However, the exam criteria require the use of lines of best fit, and many graph packages do not do this but join point to point. If a pupil has any doubts about using a graph package to draw a line of best fit, they should plot the graph by hand on graph paper.

Equipment for data-logging is supplied by Philip Harris Ltd and Griffin and George Ltd; worksheets accompany the equipment.

An informed source on the topic of data-logging is *The IT in Science book of Datalogging and Control* by Roger Frost (revised 1999), ISBN 0 9520257 1 X, and *IT in Secondary Science* by the same author (revised 1999), ISBN 0 9520257 2 8. These books are available from ASE Booksales. Further information can be obtained from **www.rogerfrost.com.**

Equipment

With lower ability pupils there is often a problem keeping the reactants apart until they are ready to start timing. Various methods are proposed in books, such as using a test-tube inside a flask (see Figure 8.12a). However, manipulating this is not easy for many pupils. A better method is to use a flat-bottomed flask with a barrier (Figure 8.12b). It may be possible to get a technician to make these. It would certainly be worthwhile.

Figure 8.12
Methods of keeping the reagents apart until timing can be started.

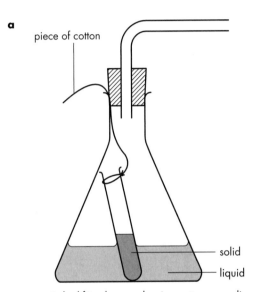

a

piece of cotton

solid
liquid

mix by lifting bung; releasing cotton, resealing and shaking

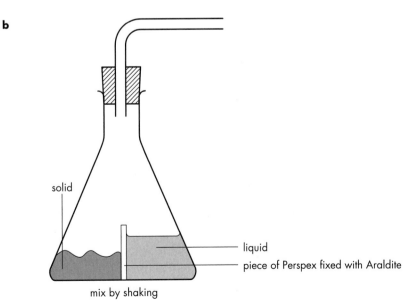

b

solid

liquid
piece of Perspex fixed with Araldite

mix by shaking

9 *Industrial chemistry*

Bob McDuell

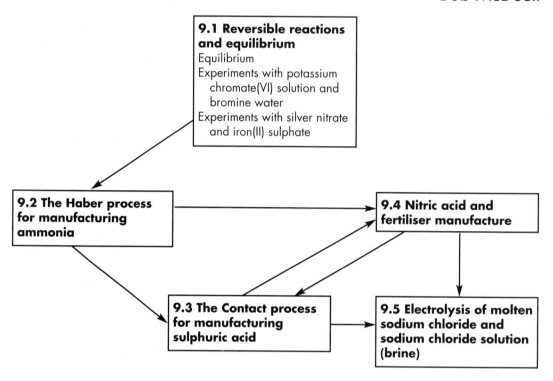

9.1 Reversible reactions and equilibrium
Equilibrium
Experiments with potassium chromate(VI) solution and bromine water
Experiments with silver nitrate and iron(II) sulphate

9.2 The Haber process for manufacturing ammonia

9.4 Nitric acid and fertiliser manufacture

9.3 The Contact process for manufacturing sulphuric acid

9.5 Electrolysis of molten sodium chloride and sodium chloride solution (brine)

♦ *Choosing a route*

In the teaching of Chemistry it is important to link the chemistry of the laboratory with the principles applied in the chemical industry when manufacturing chemicals on a large scale. Examiners are encouraged to ask questions that test the pupils' knowledge of social, economic, environmental and technological aspects of chemistry. In teaching this topic of industrial chemistry, it is important to stress that only understanding of the chemical principles is required and not details of the processes.

This section concentrates on the chemistry of the manufacture of ammonia (by the Haber process), sulphuric acid (by the Contact process), nitric acid (by oxidation of ammonia), chlorine, sodium hydroxide, hydrogen and sodium chlorate(I) (by the electrolysis of brine) and fertilisers (especially ammonium sulphate and ammonium nitrate).

Other large-scale processes are described in other parts of the course. These include:

- iron extraction and steel manufacture (Chapter 4)
- fractional distillation of crude oil (Chapter 3)
- cracking and polymerisation processes (Chapter 3)
- aluminium extraction (Chapter 4).

A study of the topic of rates of reaction (Chapter 8) should precede reversible reactions and equilibrium. Only when the concept of dynamic equilibrium is mastered should the Haber process and the Contact process be studied. The manufacture of nitric acid from ammonia follows on from the Haber process and the Contact process, and leads in to a study of fertilisers. In studying fertilisers pupils need to have studied acids, alkalis and salt production. Also, they should understand the nitrogen cycle and the role of nitrogen in plant growth. This may come from previous experience in Chemistry, but more likely from linked studies in Biology.

Pupils should have carried out or seen demonstrated experiments in electrolysis using molten electrolytes (e.g. molten zinc chloride, Chapter 4). They should be able to state the products of this electrolysis and the most able should know how to write ionic equations for the changes at the electrodes, e.g.

$$Zn^{2+} + 2e^- \rightarrow Zn$$
$$2Cl^- \rightarrow Cl_2 + 2e^-$$

Pupils will then consider electrolysis of aqueous solutions and appreciate that this often leads to more than one possible product at each electrode. They are then ready to study the industrial processes involved in the electrolysis of brine.

9.1 Reversible reactions and equilibrium

◆ *Previous knowledge and experience*

Pupils will have met examples of reversible reactions in earlier years at secondary school. For example, when zinc oxide is heated it turns from a white powder to a yellow powder. On cooling, the yellow powder turns back to the white powder. Also, blue copper(II) sulphate crystals turn white on heating to form anhydrous copper(II) sulphate. When water is added to anhydrous copper(II) sulphate, the powder turns blue, re-forming hydrated copper(II) sulphate:

$$CuSO_4.5H_2O \rightleftharpoons CuSO_4 + 5H_2O$$

Pupils understand the idea of reversible reactions with simple examples like hydrated copper(II) sulphate and zinc oxide. The concept of dynamic equilibrium, however, is more difficult to grasp. This is probably for two reasons:

- to understand it, pupils really need to think on a particle scale
- there are few good examples of dynamic equilibrium that use simple chemicals with which pupils are familiar: when examples such as bismuth(III) chloride are used, the unfamiliarity of these chemicals causes problems.

Equilibrium

In the reaction of magnesium and steam, the sign $\rightarrow$ indicates that the magnesium turns completely into magnesium oxide and that the magnesium oxide does not react again:

$$Mg + H_2O \rightarrow MgO + H_2$$

The reaction of iron and steam, however, is reversible and this is shown by the 'reversible reaction' sign:

$$3Fe(s) + 4H_2O(g) \rightleftharpoons Fe_3O_4(s) + 4H_2(g)$$

◆ *A teaching sequence*

If steam is passed over heated iron in the apparatus shown in Figure 9.1, the hydrogen gas is pushed out of the apparatus. Hence the reverse reaction cannot take place, and iron oxide and hydrogen are formed.

Figure 9.1

Heating iron in steam to produce iron oxide and hydrogen gas.

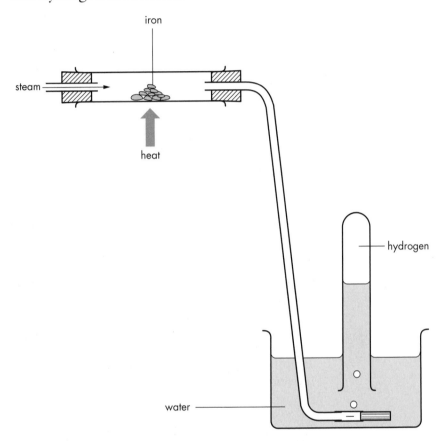

If hydrogen were passed over iron oxide in the same apparatus, the reverse reaction would take place and iron and steam would be produced. In this case the steam would be pushed out of the apparatus and the forward reaction would be prevented.

Equilibrium will only occur when there is a closed system so that nothing can escape and both the forward and reverse reactions can operate. The diagrams in Figure 9.2 (overleaf) show what happens when iron and steam are heated in a sealed container. Figure 9.2a shows the situation at the start. Iron and steam particles are present but no iron oxide or hydrogen. In Figures 9.2b, c and d the concentrations of the four substances are the same. In these cases a dynamic equilibrium has been set up.

Figure 9.2
The reaction between iron and steam would reach an equilibrium in a sealed system.

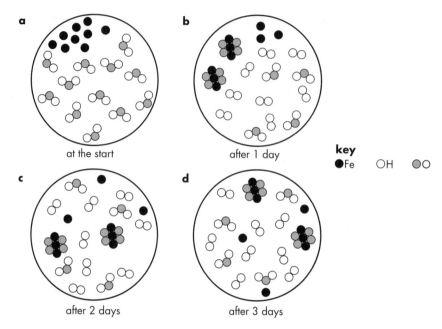

key
●Fe   OH   ◑O

It is a common misconception that at equilibrium both the forward and reverse reactions have stopped and that this is why the concentrations are unchanged. This is not the case: the concentrations are unchanged because the rate of the forward reaction is the same as the rate of the reverse reaction. Pupils may appreciate the situation if you introduce the analogy of running up a 'down' escalator: if they run up at the same speed as the escalator is moving down, they appear to be stationary.

Pupils could discuss whether any reaction is taking place in Figures 9.2b, c or d. With more able pupils you could also explain how the process could be investigated with radioactive isotopes.

Having acquired an understanding of dynamic equilibrium, pupils should then realise that the situation is very unstable and that any slight change of conditions can alter the situation. For example, going back to the escalator analogy, if the speed of the escalator were to double, the person running would start to go back down unless they started running faster. Almost certainly their position would change. The same is true with chemical equilibrium. A change in temperature, pressure or concentrations of reactants or products can alter the position of the equilibrium. If it moves to favour the formation of products, the equilibrium moves to the right. If it moves to favour the formation of reactants, the equilibrium moves to the left. (A detailed study of this is made at a higher level.)

Experiments with potassium chromate(VI) solution and bromine water

Materials

- eye protection
- 4 test-tubes, 125×16 mm
- pieces of white paper
- 2 dropping pipettes
- 2 small beakers
- bromine water
- sulphuric acid (0.1 mol/dm^3)
- sodium hydroxide (0.1 mol/dm^3)
- potassium chromate(VI) solution (0.1 mol/dm^3)

Safety

- *1% bromine water (0.06 mol/dm^3) is toxic and corrosive; 0.1% bromine water (0.006 mol/dm^3) is harmful and an irritant. Use a very dilute solution to avoid problems.*
- *0.1 mol/dm^3 sodium hydroxide is an irritant.*
- *0.1 mol/dm^3 potassium chromate(VI) is toxic.*
- *Wear eye protection when using irritant and toxic compounds.*

Procedure

1. Put 3 cm depth of bromine water into a test-tube.
2. Add 2 cm^3 of sulphuric acid to the test-tube *drop by drop*. Shake gently after each addition. Note any colour change.
3. Add 2 cm^3 of sodium hydroxide solution *drop by drop*. Shake gently after each addition. Note any colour change.
4. Repeat the experiment using potassium chromate(VI) solution in place of the bromine water. Wear eye protection.

What you might expect

With bromine water, the relevant equation is

$$Br_2(aq) \; + \; H_2O(l) \; \rightleftharpoons \; H^+(aq) \; + \; Br^-(aq) \; + \; HOBr(aq)$$
$$\text{red} \qquad\qquad\qquad\qquad \text{colourless}$$

When acid is added to bromine water the red colour remains. When alkali is added, the solution goes colourless as the equilibrium moves to the right.

With potassium chromate(VI) solution, the relevant equation is

$$2CrO_4{}^{2-}(aq) \; + \; 2H^+(aq) \; \rightleftharpoons \; Cr_2O_7{}^{2-}(aq) + H_2O(l)$$
$$\text{yellow} \qquad\qquad\qquad\qquad \text{orange}$$

When acid is added to the potassium chromate(VI) solution, the solution turns orange, owing to the formation of dichromate(VI); the equilibrium has moved to the right. When alkali is added, the solution turns yellow as the equilibrium moves to the left and chromate(VI) ions are re-formed. Pupils might suggest that an indicator acts in a similar way.

Experiments with silver nitrate and iron(II) sulphate

Materials

- eye protection
- 4 test-tubes, 125 × 16 mm
- silver nitrate solution (0.1 mol/dm³)
- iron(II) sulphate solution – this has to be freshly prepared by dissolving the solid in water, as it is readily oxidised
- iron(III) chloride solution (1 mol/dm³)
- freshly prepared potassium hexacyanoferrate(III) solution (1 mol/dm³)
- potassium thiocyanate solution (1 mol/dm³

Safety

- 0.1 mol/dm³ silver nitrate is low hazard but avoid skin contact.
- Iron(II) sulphate is a harmful solid.
- Iron(III) chloride solution is harmful above a concentration of 0.75 mol/dm³.
- Potassium hexacyanoferrate(III) solution is irritant.
- Potassium thiocyanate is a harmful solid.
- Wear eye protection when using irritant and toxic compounds.

Procedure

1. Prepare a solution of iron(II) sulphate in a test-tube and add a couple of drops of potassium hexacyanoferrate(III). Note the colour of the solution.
2. Put some iron(III) chloride into a test-tube and add a couple of drops of potassium thiocyanate solution.
3. Record the colours of the solutions from steps 1 and 2 in a table.
4. Mix together equal volumes of silver nitrate solution and freshly prepared iron(II) sulphate solution.
5. Decant off the solution from the solid that is produced. Test the solution with potassium hexacyanaferrate(II) solution and with potassium thiocyanate solution.
6. Add excess iron(III) chloride solution to the solid remaining in the test-tube. What happens to the solid? How can this be explained?

What you might expect

This table shows sample results from step 3.

Ion present	Chemical added	Colour of solution
Iron(II), Fe^{2+}	Potassium hexacyanoferrate(III)	Deep blue
Iron(III), Fe^{3+}	Potassium thiocyanate	Red

The tests for iron(II) and iron(III) using potassium hexacyanoferrate(III) and potassium thiocyanate are important for understanding what happens later in the experiment. When iron(II) sulphate and silver nitrate solutions are mixed the following reaction takes place:

$$Ag^+(aq) + Fe^{2+}(aq) \rightleftharpoons Ag(s) + Fe^{3+}(aq)$$

As the silver is precipitated, the equilibrium moves further to the right to produce iron(III) ions. Therefore the results obtained in step 5 are as follows.

Chemical added to solution	Colour of solution	Conclusion
Potassium hexacyanoferrate(III)	Not blue	No Fe^{2+} ions present
Potassium thiocyanate	Red	Fe^{3+} ions present

When iron(III) chloride is added to the solid (silver) in step 6, the equilibrium moves to the left, the silver dissolves and iron(II) ions are produced.

♦ *Enhancement ideas*

♦ Research other examples of reversible reactions.

♦ *Further activities*

♦ *Classic Chemistry Demonstrations* contains the following experiments relevant to this section:

4   The equilibrium between ICl and ICl_3
8   The equilibrium between $Co(H_2O)_6^{2+}$ and $CoCl_4^{2-}$
11 Phenolphthalein as an indicator
12 Strong and weak acids – the common ion effect
40 Equilibria involving carbon dioxide
82 The equilibrium between nitrogen dioxide and dinitrogen tetroxide
88 The equilibrium between bismuth oxide chloride and bismuth trichloride

9.2 The Haber process for manufacturing ammonia

◆ *Previous knowledge and experience*

Pupils will have no previous knowledge of processes to produce ammonia. They may be familiar with ammonia solution, which is used as a cleaning agent and is alkaline.

◆ *A teaching sequence*

The three elements necessary in large amounts for plant growth are nitrogen, phosphorus and potassium. Until the beginning of the 20th century, deposits of nitrogen fertilisers such as sodium nitrate, bird droppings and animal manure were sufficient to put nitrogen back into the soil to maintain soil fertility. With the growth in population and the demand for nitrates for explosives, a method of making ammonia synthetically was necessary. Figure 9.3 shows the growth of world population in the 20th century and the growth in the production of ammonia.

Figure 9.3
Ammonia production and the growth in world population, 1900–1980. (Source: Foundation Chemistry, 2nd edition by B. McDuell, 1983. Nelson, Walton-on-Thames.

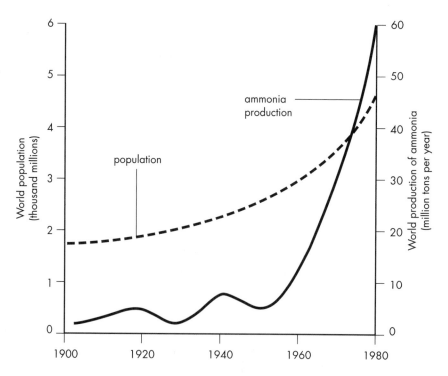

In 1908, Fritz Haber developed a process to combine nitrogen and hydrogen to produce ammonia:

$$N_2(g) + 3H_2(g) \rightleftharpoons 2NH_3(g)$$

Today the nitrogen is obtained from the fractional distillation of liquid air and the hydrogen is obtained from cracking methane or other fractions from petroleum. The reaction is reversible, so the percentage of ammonia produced depends upon the conditions. Figure 9.4 shows a diagram of the industrial plant used to produce ammonia. Only about 10% of the nitrogen and hydrogen are combined; the unreacted nitrogen and hydrogen gases are recycled. The ammonia is removed from the mixture of gases. The boiling points in Table 9.1 show that ammonia liquefies at a higher temperature than nitrogen or hydrogen.

Figure 9.4
The Haber process for manufacturing ammonia.

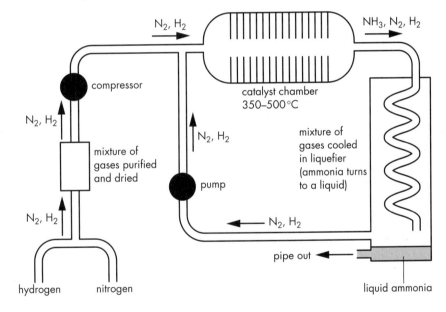

Table 9.1 *Boiling points of the gases involved in the Haber process.*

Gas	Boiling point (°C)
Nitrogen	−196
Hydrogen	−252
Ammonia	−33

Pupils should be given a flow diagram such as Figure 9.4 or the one from page 4 of *Industrial Chemistry* from the Royal Society of Chemistry, to label as fully as possible.

Figure 9.5
The equilibrium proportion of ammonia at different temperatures and pressures.

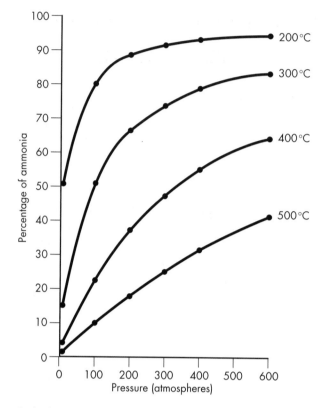

Figure 9.5 shows the percentage of ammonia produced at different temperatures and pressures under equilibrium conditions. It is important to realise that within the ammonia plant equilibrium conditions never exist: the constant flow of nitrogen and hydrogen gases will displace the mixture before it has the opportunity to establish equilibrium. However, it is still true that the highest yields of ammonia are obtained at high pressures and low temperatures. High pressures are expensive as the apparatus has to have reinforced pipework, etc. to avoid explosions, and the cost of doing this restricts the pressure used. At low temperatures the yield is high but the rate of reaction is slower. A compromise therefore has to be considered to produce the best yield in a reasonable time. The use of a catalyst of finely divided iron (with alkali as a promoter) speeds up the reactions. A typical plant today operates at 450 °C and 500 atmospheres.

◆ *Enhancement ideas*

- ◆ Fritz Haber's development of the synthesis of ammonia was extremely important; research his life and work.

9.3 The Contact process for manufacturing sulphuric acid

◆ *Previous knowledge and experience*

Pupils will have used dilute sulphuric acid. They will have no knowledge of the method used in its manufacture.

◆ *A teaching sequence*

The manufacture of sulphuric acid by the Contact process is not usually in GCSE Science schemes but is included in many GCSE Chemistry courses. It is a good example of the use of a reversible reaction and the control of equilibrium conditions to achieve good yields.

The Contact process occurs in three distinct stages:

1. Sulphur (or minerals containing sulphur) are heated in air to produce sulphur(IV) oxide (sulphur dioxide):

$$S \ + \ O_2 \ \rightarrow \ SO_2$$
sulphur + oxygen → sulphur(IV) oxide

or

$$4FeS_2 \ + \ 11O_2 \rightarrow \ 2Fe_2O_3 \ + \ 8SO_2$$
iron disulphide + oxygen → iron(III) oxide + sulphur(IV) oxide

The sulphur(IV) oxide is purified at this stage because impurities, such as arsenic, would poison the catalyst at stage 2.

2. The mixture of sulphur(IV) oxide and air is passed over a heated catalyst of vanadium(V) oxide.

$$2SO_2 \ + \ O_2 \ \rightleftharpoons \ 2SO_3$$
sulphur(IV) oxide + oxygen ⇌ sulphur(VI) oxide

As this reaction is reversible, conditions will affect the yield and hence the final yield of sulphuric acid. The forward reaction is exothermic, and the best yield of sulphur(VI) oxide is obtained at low temperatures. However, at low temperatures reactions are very slow and, although a high yield is obtained, the time taken makes it uneconomic. A compromise is made between getting a high yield and carrying out the process quickly. The temperature used is about 450 °C and the catalyst speeds up the reactions.

In theory, a higher yield would be obtained at a higher pressure. However, the yield is good without increasing the pressure, and the extra costs of operating at higher pressures would not be recovered.

3. In theory, all that has to be done is to dissolve the sulphur(VI) oxide in water to form sulphuric acid:

$$SO_3 + H_2O \rightarrow H_2SO_4$$

But this process is very exothermic and the sulphuric acid boils, producing sulphuric acid vapour which is harmful. A better method is to dissolve the sulphur(VI) oxide in concentrated sulphuric acid to form oleum (fuming sulphuric acid). This forms concentrated sulphuric acid when diluted with water:

$$SO_3 \quad + \quad H_2SO_4 \quad \rightarrow \quad H_2S_2O_7$$
$$\text{sulphur(VI) oxide} + \text{sulphuric acid} \rightarrow \quad \text{oleum}$$

$$H_2S_2O_7 \quad + \quad H_2O \quad \rightarrow \quad 2H_2SO_4$$
$$\text{oleum} \quad + \quad \text{water} \quad \rightarrow \text{sulphuric acid}$$

Figure 9.6 summaries the Contact process.

Figure 9.6
The Contact process for manufacturing sulphuric acid.

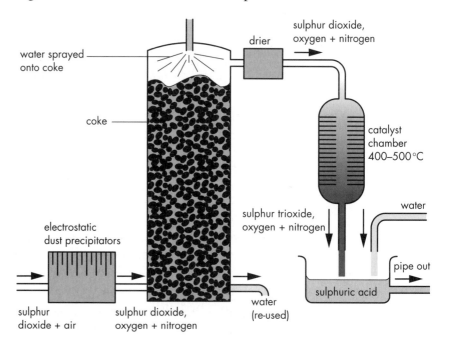

Pupils should fully label a flow diagram of the process (for example, Figure 9.6 or the one from page 12 of *Industrial Chemistry* from the Royal Society of Chemistry).

9.4 Nitric acid and fertiliser manufacture

◆ *Previous knowledge and experience*

Pupils will have used dilute nitric acid. They will have no experience of the method used to manufacture nitric acid. Pupils may have some experience of fertilisers, especially from Biology. Nitrogen, potassium and phosphorus are absorbed in solution through the roots of a plant. Other elements, called 'trace elements', are required in small amounts.

◆ *A teaching sequence*

Nitric acid is manufactured by the oxidation of ammonia. Ammonia does not burn in air but it does burn in oxygen.

Demonstration of burning ammonia in oxygen
Materials
- eye protection
- glass tube, approximately 4 cm in diameter and 15–20 cm long, with a cork carrying 2 pieces of glass tubing (Figure 9.7, overleaf)
- splints
- litmus paper
- Bunsen burner
- heatproof mat
- connecting tube
- glass rod
- 2 stands, bosses and clamps
- access to oxygen cylinder
- ammonia generator containing 0.880 ammonia (concentrated ammonia, with a specific gravity of 0.880)

Safety
- *Oxygen is very oxidising.*
- *Ammonia is toxic; its vapour is irritating to the eyes and the respiratory system. A concentrated solution of ammonia is corrosive and causes burns.*
- *Potassium hydroxide pellets are corrosive.*
- *Glass wool is an irritant.*
- *Wear eye protection when using a Bunsen burner and using ammonia.*

Figure 9.7
*Burning ammonia
in oxygen.*

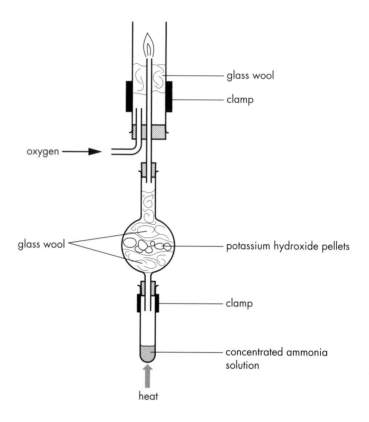

Labels in figure:
- glass wool
- clamp
- oxygen →
- glass wool
- potassium hydroxide pellets
- clamp
- concentrated ammonia solution
- heat

Procedure
1. Set up the apparatus as shown in Figure 9.7.
2. Turn on the oxygen supply to provide a gentle stream of oxygen.
3. Use a glowing splint to show the presence of oxygen in the glass tube.
4. Heat the ammonia generator gently to produce ammonia, and use damp red litmus paper to show the presence of ammonia in the glass tube.
5. Apply a lighted splint to the ammonia jet.

What you might expect
The ammonia starts to burn with a greenish flame. The reaction taking place is

$$4NH_3(g) + 3O_2(g) \rightarrow 2N_2(g) + 6H_2O(g)$$
ammonia + oxygen → nitrogen + steam

This reaction is not useful in converting ammonia into nitric acid as nitrogen is very unreactive. A more useful reaction is the reaction of ammonia with oxygen in the present of a platinum catalyst. This produces nitrogen monoxide, which reacts with oxygen to form brown nitrogen dioxide.

Demonstration of the catalytic oxidation of ammonia

Materials

- eye protection
- 250 cm^3 beaker
- piece of platinum wire
- tongs
- Bunsen burner
- jet of air or oxygen
- 0.880 ammonia solution

Safety

- *0.880 ammonia solution is corrosive.*
- *Carry out the demonstration in a fume cupboard.*
- *Wear eye protection when using a Bunsen burner and when using ammonia.*

Procedure

Figure 9.8
Reacting ammonia and oxygen in the presence of a platinum catalyst.

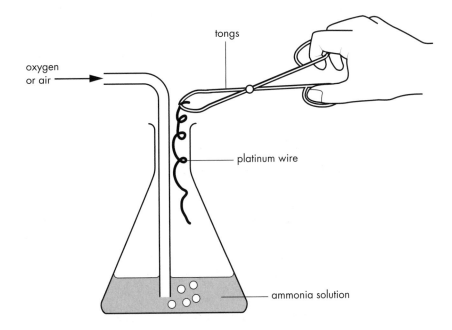

tongs

oxygen or air →

platinum wire

ammonia solution

1. Put about 3 cm depth of 0.880 ammonia solution into the beaker (see Figure 9.8)
2. Allow a steady steam of oxygen or air to bubble through the solution.
3. Coil the platinum wire into a spiral.
4. Hold the wire in tongs and heat it in a hot Bunsen burner flame until it is red-hot.
5. Lower it quickly into the flask.

<u>What you might expect</u>
The platinum wire should start to glow more strongly as the reaction seen is exothermic. You can also see brown fumes being formed (these show up more clearly against a white background).

There are two reactions taking place:

$$4NH_3(g) \ + \ 5O_2(g) \ \rightarrow \ 4NO(g) \ + \ 6H_2O(g)$$

ammonia + oxygen → nitrogen monoxide + steam

$$2NO(g) \ + \ O_2(g) \ \rightleftharpoons \ 2NO_2(g)$$

nitrogen monoxide + oxygen ⇌ nitrogen dioxide

If nitrogen dioxide is dissolved in water in the presence of air, nitric acid is formed.

These reactions are the ones used in industry to convert ammonia into nitric acid. In the industrial process (summarised in Figure 9.9), a mixture of 10% ammonia and 90% air is passed over a heated platinum gauze catalyst.

Figure 9.9
The industrial process for making nitric acid.

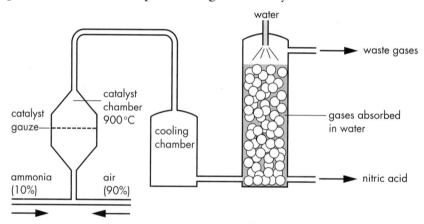

An exothermic reaction takes place:

$$4NH_3(g) \ + \ 5O_2(g) \ \rightarrow \ 4NO(g) \ + \ 6H_2O(g)$$

ammonia + oxygen → nitrogen monoxide + steam

When the mixture of gases is cooled a further reaction occurs:

$$2NO(g) \ + \ O_2(g) \ \rightleftharpoons \ 2NO_2(g)$$

nitrogen monoxide + oxygen ⇌ nitrogen dioxide

The gases are then dissolved in water to form nitric acid:

$$4NO_2(g) \ + \ 2H_2O(l) \ + \ O_2(g) \ \rightarrow \ 4HNO_3(l)$$

nitrogen dioxide + water + oxygen → nitric acid

The nitric acid produced in this process is used for making fertilisers, explosives and nylon.

Preparation of ammonium sulphate

Ammonium sulphate is widely used as a fertiliser. It is made here rather than ammonium nitrate because ammonium nitrate is unstable and can cause explosions.

Materials
- eye protection
- evaporating basin
- glass rod
- Bunsen burner
- heatproof mat, tripod and gauze
- measuring cylinder (25 cm^3)
- filter flask, funnel, filter paper and pump
- sulphuric acid (1 mol/dm^3)
- ammonia solution (2 mol/dm^3)

Safety
- *1 mol/dm^3 sulphuric acid is an irritant.*
- *It is essential to carry out this experiment in a well-ventilated laboratory.*
- *Wear eye protection throughout.*

Procedure
1. Put about 20 cm^3 of sulphuric acid into an evaporating basin.
2. Add ammonia solution a little at a time until the solution has a definite smell of ammonia.
3. Evaporate the solution until only about one-fifth of the original volume remains. Leave the solution to cool.
4. Filter off the crystals and dry them with filter paper.

What you should expect
Usually in salt preparations (see page 38), an indicator is used to detect when exactly the correct quantities of the acid and the alkali are present. Here it is not necessary, provided that the ammonia (the alkali) is in excess: the excess ammonia will boil off on evaporation. Colourless crystals of ammonium sulphate are formed.

◆ *Enhancement ideas*

- ◆ Carry out experiments to grow plants with deficiencies in nitrogen, potassium and phosphorus. This could be done in conjunction with work in Biology.
- ◆ Fertilisers are often given 'NPK' values. Look at the compositions of different fertilisers.

9.5 Electrolysis of molten sodium chloride and sodium chloride solution (brine)

◆ *Previous knowledge and experience*

In Chapter 4, electrolysis was introduced as a means of splitting up stable compounds using electricity; electrolysis of molten zinc chloride produces zinc and chlorine and electrolysis of copper(II) sulphate solution produces copper at the negative electrode. In this section we will consider the electrolysis of molten sodium chloride and sodium chloride solution (called 'brine'). These are very important industrial processes.

◆ *A teaching sequence*

Electrolysis of molten sodium chloride

Like zinc chloride, sodium chloride is an electrolyte. It is not split up by electricity when it is solid but is when it is molten or dissolved in water (in aqueous solution).

Sodium chloride is made up of ions – Na^+ ions and Cl^- ions. In the solid they are regularly arranged in the lattice and are not free to move. When sodium chloride is melted or dissolved in water the ions are free to move. It is the ions that carry the charge through the melt or through the solution. Pupils frequently compare electrical conductivity in a melt or solution with electrical conductivity in a metal. They write that electricity passes through a melt or solution by passage of free electrons. This is untrue: it is due to the passage of ions.

Electrolysis of molten sodium chloride is not easy to demonstrate in the laboratory, owing to the high melting point of sodium chloride.

In industry, molten sodium chloride is electrolysed in the Downs cell (Figure 9.10). The products are sodium metal and chlorine gas. Calcium chloride is added to the sodium chloride in the cell to lower the melting point of the solid from $800\,°C$ to $600\,°C$; this reduces fuel costs.

Figure 9.10
The Downs cell for extracting sodium. (Adapted from: GCSE Science classbook by D. Baylis, G. Booth & B. McDuell, 1996. Letts Educational, London.)

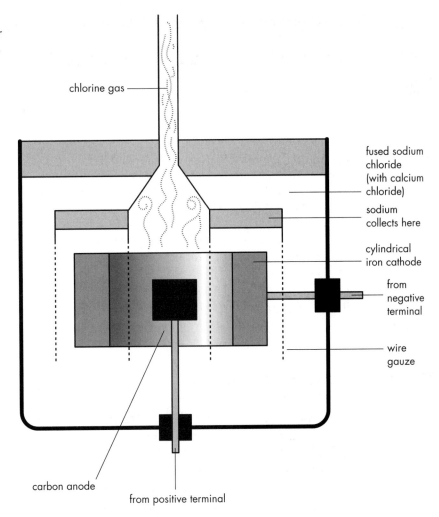

chlorine gas

fused sodium chloride (with calcium chloride)

sodium collects here

cylindrical iron cathode

from negative terminal

wire gauze

carbon anode

from positive terminal

The reaction at the cathode is

$$Na^+ + e^- \rightarrow Na$$

and at the anode:

$$2Cl^- \rightarrow Cl_2 + 2e^-$$

The low density of the sodium means that it floats up, to be collected in an inverted trough, while chlorine is collected by a hood.

It is important to emphasise that the sodium and chlorine must be kept apart once produced, or they would react to give sodium chloride again. Also, no water must be present or the sodium would react with it.

This process is important for producing sodium.

Electrolysis of sodium chloride solution

Materials

- eye protection
- electrolysis cell
- 2 cables with crocodile clips
- 9–12 V power pack
- stand, boss and clamp
- small rimless test-tubes
- sodium chloride solution (1 mol/dm³)
- blue litmus paper
- splints
- Bunsen burner
- universal indicator solution

Safety

- *Chlorine is a toxic gas with a choking smell.*
- *Warn asthmatics not to inhale gases.*
- *This experiment should be carried out in a fume cupboard or a well-ventilated laboratory. If it is carried out in an open lab, do not allow the electrolysis to run for longer than necessary to detect the products.*
- *Wear eye protection.*

Procedure

Figure 9.11
Collecting the gases produced at the electrodes during the electrolysis of brine.

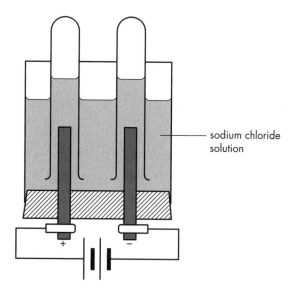

sodium chloride solution

1. Set up the apparatus as shown in Figure 9.11.
2. Collect the gases produced at the electrodes.
3. Test the gases with a lighted splint and with damp blue litmus paper.
4. Repeat the experiment with a few drops of universal indicator added to the solution.

Electrolysis of molten sodium chloride and sodium chloride solution (brine)

<u>What you might expect</u>
You should collect gases in the two test-tubes. The gas collected in the test-tube above the positive electrode should turn damp blue litmus paper red and then turn it white; this is chlorine. The gas collected in the test-tube above the negative electrode burns with a squeaky pop when a lighted splint is added; it is hydrogen.

At the anode: $2Cl^- \rightarrow Cl_2 + 2e^-$
At the cathode: $2H^+ + 2e^- \rightarrow H_2$

During the electrolysis hydrogen ions are discharged. You should see the solution turn purple when Universal indicator is added. The colour will first be seen around the negative electrode, where there will be an excess of OH^- ions.

Electrolysis of sodium chloride solution (brine) in industry

Chlorine and sodium hydroxide are very important industrial chemicals. The annual production of chlorine in the UK is 1.5 million tonnes and that of sodium hydroxide is 1.2 million tonnes.

Three methods are used for the electrolysis of sodium chloride solution; these are compared in Table 9.2.

Table 9.2   *Methods for the electrolysis of sodium chloride.*

Cell type	Construction	Operation of cell	Quality of sodium hydroxide produced
Diaphragm cell	Relatively simple and inexpensive	• Frequent replacement of diaphragm (every 100 days) • Constant current load needed • Operates at 3.8 V	Must be evaporated to concentrate to 50% and crystallise out salt
Mercury cell	• Expensive to construct and charge with mercury • Needs large floor area	• Mercury is hazardous and must be reclaimed from effluent • Can operate over a wide current load • Operates at 4.5 V	• High purity • Produced at 50% concentration
Membrane cell	• Cheap to construct and install • Requires high quanlity brine (low in calcium and magnesium)	• Can operate at varying current loads • Operates at 3.3 V • Membrane changed every 2–3 years	• High purity • Must be evaporated to 50% and crystallise out salt

It is essential in all three processes that the chlorine and the sodium hydroxide solution are kept separate. If they mix, sodium chlorate(I) (sodium hypochlorite) is produced (this is the essential ingredient in household bleaches):

$$Cl_2(g) + 2NaOH(aq) \rightarrow NaOCl(aq) + NaCl(aq) + H_2O(l)$$

◆ Enhancement ideas

- ◆ Research the three methods used for the industrial electrolysis of sodium chloride solution.
- ◆ Sodium chloride is also used in the Solvay process to produce sodium carbonate and sodium hydrogencarbonate. Find out the details of this process.

◆ Further activities

- ◆ *Classic Chemistry Demonstrations* contains the following demonstration which is relevant to this section:
  44 Movement of ions during electrolysis

◆ References

Classic Chemistry Demonstrations (1998). Royal Society of Chemistry. ISBN 1 870343 28 7. This book was distributed free to all schools in the UK. Further information is available from Royal Society of Chemistry, Burlington House, Piccadilly, London W1V 0BN.

◆ Other resources

The equilibrium of the cobalt chloride–water system can be studied on a small scale using the instructions in *Microscale Chemistry*, No. 18, Royal Society of Chemistry, 1998, ISBN 1 870343 49 2.

The Essential Chemical Industry, from the Chemical Industry Education Centre, Department of Chemistry, University of York, York YO1 5DD, gives up-to-date statistics of the chemical industry (new edition 1999).

The Chemical Industries Association, King Buildings, Smith
Square, London SW1P 3JJ, produces a booklet entitled
The UK Chemical Industry annually.

Sections A22 'Ammonia, fertilisers and food production' and
B14 'Chemistry and the world food problem' of Revised
Nuffield Chemistry *Teachers' Guide II* (Longman, 1978,
ISBN 0 582 04636 X) are very useful for teaching this
section.

Videos

For information about all major industrial processes, the video
Industrial Chemistry for Schools and Colleges, produced by
the Royal Society of Chemistry, is very good. A copy was
distributed to schools and colleges in 1999. Further copies
are available from The Royal Society of Chemistry, Turpin
Distribution Services Ltd, Blackhorse Road, Letchworth,
Herts SG6 1HN. This video is supported by a booklet
(ISBN 0 85404 945 2) that includes answers to some
questions students might ask and some questions for
assessing student understanding of the processes. The
relevant section of the video should be shown after the
chemistry of the process has been taught. Each section lasts
about 5 minutes. The clips describe how each process works
and look at the raw materials, the uses of the products and
the geographical siting of the plant. The aim is to give pupils
an idea of the size and scale of the plant and to give them
some of the feel of a site visit.

Chemical Equilibrium produced by Classroom Video, Darby
House, Bletchingley Road, Merstham, Redhill, Surrey
RH1 3DN, is suitable for pupils aged 15–18.

Visits

Catalyst, The Museum of the Chemical Industry, Gossage
Building, Mersey Road, Widnes WA8 0DF.

The Chemical Industry Gallery of the Science Museum,
Exhibition Road, London SW7 2DD.

The Salt Museum, 162 London Road, Northwich, Cheshire
CW9 8AB.

◆ *Useful addresses*

Companies producing material useful for teachers include the following:

The Aluminium Federation, Broadway House, Calthorpe Road, Birmingham B15 1TN.

The Associated Octel Company Ltd, PO Box 17, Oil Sites Road, Ellesmere Port, South Wirral L65 4HF.

BOC Gases, Corporate Relations Department, 10 Priestley Road, Surrey Research Park, Guildford, Surrey GU2 5XY.

BP Educational Services (BPES), PO Box 934, Poole, Dorset BH17 7BR.

British Steel Educational Service, PO Box 10, Wetherby, W Yorks LS23 7EL.

Copper Development Association, Verulam Industrial Estate, 224 London Road, St Albans, Herts AL1 1AQ.

ICI Chlor-Chemicals, Room B111, HQ Main Building, PO Box 114, The Heath, Runcorn, Cheshire WA7 4QG.

Rio-Tinto plc (Educational Resources Service), 6 St James's Square, London SW1Y 4LD.

Shell Education Service, PO Box 46, Newbury, Berks RG14 2YX.

Appendix

The following two pages contain a photocopiable example of a student form for assessing risk, and a completed example. The student form is designed to encourage pupils to think about the dangers involved in an experiment and to plan a strategy for working safely.

Student Form for Assessing Risk

Proposed practical activity ....

....

Name(s) of pupil(s) completing form ....

Class/set .... Date ....

Hazardous chemical or microorganism being used or made, or hazardous procedure or equipment	Nature of the hazard(s)	Source(s) of information	Control measures to reduce the risks

Checked by .... Date ....

Proposed practical activity .Preparation.of.copper(II).sulphate.from.copper(II).oxide................

....

Name(s) of pupil(s) completing form ..A.K.Smith...

Class/set ..11A... Date .16.11.99...................................

Hazardous chemical or microorganism being used or made, or hazardous procedure or equipment	Nature of the hazard(s)	Source(s) of information	Control measures to reduce the risks
Sulphuric acid	Acid is corrosive if 1.5 mol/dm³ or more. It is an irritant if 0.5 mol/dm³ or more	Bottle label CLEAPSS Student Safety Sheets	Use lowest possible concentration – 0.5 mol/dm³. Wear eye protection
Copper(II) oxide	Solid is harmful if swallowed. Dust irritates eyes and lungs. Heating copper(II) oxide and sulphuric acid – solution may boil over. Hot tripods	Bottle label CLEAPSS Student Safety Sheet Teacher warning Past experience	Wash hands thoroughly after use. Wear eye protection Control Bunsen burner flame Pay attention
Copper(II) sulphate	Solid and solutions more concentrated than 1 mol/dm³ are harmful	CLEAPSS Student Safety Sheets	Wash hands after use
Evaporating solution to form saturated solution.	Solution may boil over	Past experience	Turn down gas supply as liquid boils. Wear eye protection

Checked by ...GR...................................... Date .17.11.99...........................
(Adapted with permission from CLEAPSS)

Index